OPTICAL METROLOGY

Second Edition

OPTICAL METROLOGY

Second Edition

Kjell J. Gåsvik

Spectra Vision AS, Trondheim, Norway

JOHN WILEY & SONS

Chichester · New York · Brisbane · Toronto · Singapore

National   01243 779777
International (+44) 1243 779777

Other Wiley Editorial Offices

John Wiley & Sons, Inc., 605 Third Avenue,
New York, NY 10158-0012, USA

Jacaranda Wiley Ltd, 33 Park Road, Milton,
Queensland 4064, Australia

John Wiley & Sons (Canada) Ltd, 22 Worcester Road,
Rexdale, Ontario M9W 1L1, Canada

John Wiley & Sons (SEA) Pte Ltd, 37 Jalan Pemimpin #05-04,
Block B, Union Industrial Building, Singapore 2057

Library of Congress Cataloging-in-Publication Data

Gåsvik, Kjell J.
    Optical metrology / Kjell J. Gåsvik. — 2nd ed.
        p.   cm.
    ISBN 0 471 95474 8. — ISBN 0 471 95528 0
    1. Optical measurements—Congresses.   I. Title.
QC367.G37   1995   95-4095
621.36—dc20   CIP

British Library Cataloging in Publication Data

A catalogue record for this book is available from the British Library

ISBN 0471 95474 8 (PPc)   ISBN 0 471 95528 0 (PR)

Typeset in 10/12pt Times by Thomson Press (India) Ltd, New Delhi
Printed and bound in Great Britain by Bookcraft (Bath) Ltd
This book is printed on acid-free paper responsibly manufactured from sustainable forestation,
for which at least two trees are planted for each one used for paper production.

Contents

Preface to the second edition

In this edition of *Optical Metrology*, four new main chapter have been written. These are:

- Chapter 2, Gaussian optics with an introduction to the matrix method of calculating the performance of optical imaging systems. I believe this will fill a need in introductory courses in optics.
- Chapter 5, Light sources and detectors. Here an extensive treatment of lasers and laser principles is given. This treatment covers most of the laser types commercially available today. To discuss light sources in general, an introduction to radiometry and photometry is included. As the most common detector in optical metrology applications, the CCD camera is given an extensive description. Also a section on sampling theory is included.
- Chapter 10, Digital image processing. As a consequence of the ever-increasing use of digital images in optical metrology, a separate chapter on digital image processing is given. Here the frame grabber is introduced and a description of basic digital image processing techniques like the detection of edges with sub-pixel accuracy is given.
- Chapter 11, Fringe analysis. Here the many modern aspects of digital fringe-pattern measurements techniques are described, with special emphasis on phase measurement interferometry and phase unwrapping. Thus, this chapter gives the state of the art of optical metrology.

Also new sections on spatial and temporal coherence, heterodyne detection, the optical transfer function, the Wiener spectrum, shearography, refraction at an interface and fibre optics are included. In addition minor changes and amendments to the first edition are given.

Last but not least, a comprehensive number of problems are given at the end of each chapter.

I believe this comprehensive expansion of the text will make the book attractive as a textbook for courses in optics also outside the experimental mechanics society. This edition has become a bit more advanced, but the treatment is always kept as simple as possible. The trend in optical metrology seems to be more signal theory and less optics. I think this new edition reflects this trend.

It gives me great pleasure to acknowledge the many stimulating discussions and invaluable comments on the new manuscript by Professor H. M. Pedersen at the Department of Physics, The Norwegian Institute of Technology. In spite of his expert advice, inevitably errors will be found and these are solely my responsibility.

1 Basics

1.1 WAVE MOTION. THE ELECTROMAGNETIC SPECTRUM

Figure 1.1 shows a snapshot of an harmonic wave that propagates in the z-direction. The disturbance $\psi(z, t)$ is given as

$$\psi(z, t) = U \cos\left[2\pi\left(\frac{z}{\lambda} - vt\right) + \delta \right] \tag{1.1}$$

The argument of the cosine function is termed the phase and δ the phase constant. Other parameters involved are

U = the amplitude
λ = the wavelength
v = the frequency (the number of waves per unit time)
$k = 2\pi/\lambda$ the wave number

The relation between the frequency and the wavelength is given by

$$\lambda v = v \tag{1.2}$$

where

v = the wave velocity

$\psi(z, t)$ might represent the field in an electromagnetic wave for which we have

$$v = c = 3 \times 10^8 \, \text{m/s}$$

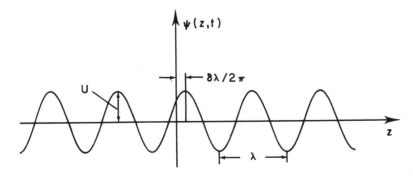

Fig. 1.1. *Harmonic wave*

Table 1.1. *The electromagnetic spectrum (From Young (1968))*

f Frequency, cycles/sec	λ Wavelength, meters	E Photon energy, electron volts				
10^2	10^6	10^{-12}	Power transmission			
10^4	10^4	10^{-10}	A M radio			
10^6	10^2	10^{-8}				
10^8	1	10^{-6}	FM radio, television			
10^{10}	10^{-2}	10^{-4}	Microwaves, radar		λ, Å	
10^{12}	10^{-4}	10^{-2}	Infrared	Red	6600	
				Orange	6100	
10^{14}	10^{-6}	1	Visible light	Yellow	5700	
10^{16}	10^{-8}	10^2	Ultraviolet	Green	5200	
10^{18}	10^{-10}	10^4	X-rays	Blue	4700	
				Violet	4,100	
10^{20}	10^{-12}	10^6	Gamma rays, cosmic rays			
10^{22}	10^{-14}	10^8				
10^{24}	10^{-16}	10^{10}				
10^{26}	10^{-18}	10^{12}	T			

The ratio of the speed c of an electromagnetic wave in vacuum to the speed v in a medium is known as the absolute index of refraction n of that medium

$$n = \frac{c}{v} \tag{1.3}$$

The electromagnetic spectrum is given in Table 1.1.

Although it does not really affect our argument, we shall mainly be concerned with visible light where

$\lambda = 400$–700 nm $(1 \text{ nm} = 10^{-9} \text{ m})$
$v = (4.3$–$7.5) \times 10^{14} \text{ Hz}$

1.2 THE PLANE WAVE. LIGHT RAYS

Electromagnetic waves are not two dimensional as in Figure 1.1, but rather three-dimensional waves. The simplest example of such waves is given in Figure 1.2 where a plane wave that propagates in the direction of the **k**-vector is sketched. Points of equal phase lie on parallel planes that are perpendicular to the propagation direction. Such

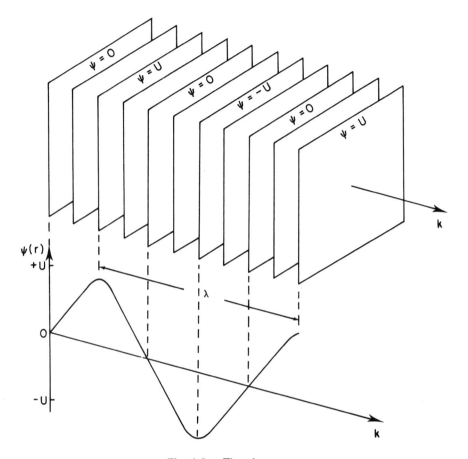

Fig. 1.2. *The plane wave*

planes are called phase planes or phase fronts. In the figure, only some of the infinite number of phase planes are drawn. Ideally, they should also have infinite extent.

Equation (1.1) describes a plane wave that propagates in the z-direction. ($z = $ constant gives equal phase for all x, y, i.e. planes that are normal to the z-direction.) In the general case where a plane wave propagates in the direction of a unit vector $\boldsymbol{n}$, the expression describing the field at an arbitrary point with radius vector $\mathbf{r} = (x, y, z)$ is given by

$$\psi(x, y, z, t) = U \cos\left[k\boldsymbol{n}\cdot\mathbf{r} - 2\pi vt + \delta\right] \tag{1.4}$$

That the scalar product fulfilling the condition $\boldsymbol{n}\cdot\mathbf{r} = $ constant describes a plane which is perpendicular to $\boldsymbol{n}$ is shown in the two-dimensional case in Figure 1.3. That this is correct also in the three-dimensional case is easily proved.

Next we give the definition of light rays. They are directed lines that are everywhere perpendicular to the phase planes. This is illustrated in Figure 1.4 where the cross-section of a rather complicated wavefront is sketched and where some of the light rays perpendicular to the wavefront are drawn.

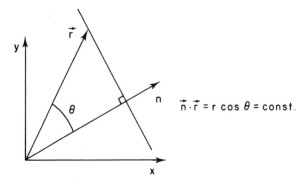

$\vec{n} \cdot \vec{r} = r \cos \theta = \text{const.}$

Fig. 1.3

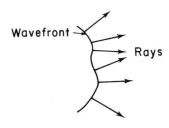

Fig. 1.4

1.3 PHASE DIFFERENCE

Let us for a moment turn back to the plane wave described by Equation (1.1). At two points z_1 and z_2 along the propagation direction, the phases are $\phi_1 = kz_1 - 2\pi vt + \delta$ and $\phi_2 = kz_2 - 2\pi vt + \delta$ respectively, and the phase difference

$$\Delta\phi = \phi_1 - \phi_2 = k(z_1 - z_2) \tag{1.5}$$

Hence, we see that the phase difference between two points along the propagation direction of a plane wave is equal to the geometrical path-length difference multiplied by the wave number. This is generally true for any light ray. When the light passes a medium different from air (vacuum), we have to multiply by the refractive index n of the medium, such that

optical path length $= n \times$ (geometrical path length)
phase difference $= k \times$ (optical path length)

1.4 COMPLEX NOTATION. COMPLEX AMPLITUDE

The expression in Equation (1.4) can be written in complex form as

$$\psi(x, y, z, t) = \text{Re}\{U e^{i(\phi - 2\pi vt)}\} \tag{1.6a}$$

where

$$\phi = k\mathbf{n}\cdot\mathbf{r} + \delta \qquad (1.6b)$$

is the spatial dependent phase. In Appendix A, some simple arithmetic rules for complex numbers are given.

In the description of wave phenomena, the notation of Equation (1.6) is commonly adopted and 'Re' is omitted because it is silently understood that the field is described by the real part.

One advantage of such complex representation of the field is that the spatial and temporal parts factorize:

$$\psi(x, y, z, t) = Ue^{i(\phi - 2\pi vt)} = Ue^{i\phi}e^{-i2\pi vt} \qquad (1.7)$$

In optical metrology (and in other branches of optics) one is most often interested in the spatial distribution of the field. Since the temporal-dependent part is known for each frequency component, we therefore can omit the factor $e^{-i2\pi vt}$ and only consider the spatial complex amplitude

$$u = Ue^{i\phi} \qquad (1.8)$$

This expression describes not only a plane wave, but a general three-dimensional wave where both the amplitude U and the phase ϕ may be functions of x, y and z.

Figure 1.5(a, b) shows examples of a cylindrical wave and a spherical wave, while in Figure 1.5(c) a more complicated wavefront resulting from reflection from a rough surface is sketched. Note that far away from the point source in Figure 1.5(b), the spherical wave is nearly a plane wave over a small area. A point source at infinity, represents a plane wave.

1.5 OBLIQUE INCIDENCE OF A PLANE WAVE

In optics, one is often interested in the amplitude and phase distribution of a wave over fixed planes in space. Let us consider the simple case sketched in Figure 1.6 where a plane wave falls obliquely on to a plane parallel to the xz-plane a distance z from it. The wave propagates along the unit vector $\mathbf{n}$ which is lying in the xz-plane (defined as the plane of incidence) and makes an angle θ to the z-axis. The components of the $\mathbf{n}$- and $\mathbf{r}$-vectors are therefore

$$\mathbf{n} = (\sin\theta, 0, \cos\theta)$$
$$\mathbf{r} = (x, y, z)$$

These expressions put into Equation (1.6) (Re and temporal part omitted) give

$$u = Ue^{ik(x\sin\theta + z\cos\theta)} \qquad (1.9a)$$

For $z = 0$ (the xy-plane) this reduces to

$$u = Ue^{ikx\sin\theta} \qquad (1.9b)$$

(a)

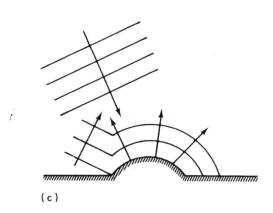

(b)

(c)

Fig. 1.5 *((a) and (b) from Hecht & Zajac (1974), Figures 2.16 and 2.17. Reprinted with permission.)*

1.6   THE SPHERICAL WAVE

A spherical wave, illustrated in Figure 1.5(b), is a wave emitted by a point source. It should be easily realized that the complex amplitude representing a spherical wave must be of the form

$$u = \frac{U}{r} e^{ikr} \tag{1.10}$$

Fig. 1.6

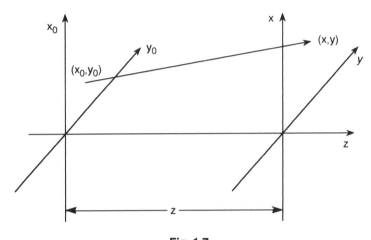

Fig. 1.7

where r is the radial distance from the point source. We see that the phase of this wave is constant for $r =$ constant, i.e. the phase fronts are spheres centred at the point source. The r in the denominator of Equation (1.10) expresses the fact that the amplitude decreases as the inverse of the distance from the point source.

Consider Figure 1.7 where a point source is lying in the x_0, y_0-plane at a point of coordinates x_0, y_0. The field amplitude in a plane parallel to the $x_0 y_0$-plane at a distance z then will be given by Equation (1.10) with

$$r = \sqrt{z^2 + (x - x_0)^2 + (y - y_0)^2} \tag{1.11}$$

where x, y are the coordinates of the illuminated plane. This expression is, however, rather cumbersome to work with. One therefore usually makes some approximations, the first of which is to replace z for r in the denominator of Equation (1.10). This approximation cannot be put into the exponent since the resulting error is multiplied by the very large number k. A convenient means for approximation of the phase is offered by a binomial expansion of the square root, viz.

$$r = z\sqrt{1 + \left(\frac{x - x_0}{z}\right)^2 + \left(\frac{y - y_0}{z}\right)^2} \approx z\left[1 + \frac{1}{2}\left(\frac{x - x_0}{z}\right)^2 + \frac{1}{2}\left(\frac{y - y_0}{z}\right)^2\right] \quad (1.12)$$

where r is approximated by the two first terms of the expansion.

The complex field amplitude in the xy-plane resulting from a point source at x_0, y_0 in the $x_0 y_0$-plane is therefore given by

$$u(x, y, z) = \frac{U}{z} e^{ikz} e^{i(k/2z)[(x - x_0)^2 + (y - y_0)^2]} \quad (1.13)$$

The approximations leading to this expression are called the Fresnel approximations. We shall here not discuss the detailed conditions for its validity, but it is clear that $(x - x_0)$ and $(y - y_0)$ must be much less than the distance z.

1.7   THE INTENSITY

With regard to the registration of light, we are faced with the fact that media for direct recording of the field amplitude do not exist. The most common detectors (like the eye, photodiodes, multiplication tubes, photographic film, etc.) register the irradiance (i.e. effect per unit area) which is proportional to the field amplitude absolutely squared:

$$I = |u|^2 = U^2 \quad (1.14)$$

This important quantity will hereafter be called the intensity.

We mention that the correct relation between U^2 and the irradiance is given by

$$I = \frac{\varepsilon v}{2} U^2 \quad (1.15)$$

where v is the wave velocity and ε is known as the electric permittivity of the medium. In this book, we will need this relation only when calculating the transmittance at an interface (see Section 9.4).

1.8   GEOMETRICAL OPTICS

For completeness, we refer to the three laws of geometrical optics:

(1) Rectilinear propagation in a uniform, homogeneous medium.
(2) Reflection. On reflection from a mirror, the angle of reflection is equal to the angle of incidence (see Figure 1.8). In this context we mention that on reflection (scattering) from a rough surface (roughness $> \lambda$) the light will be scattered in all directions (see Figure 1.9).
(3) Refraction. When light propagates from a medium of refractive index n_1 into a medium of refractive index n_2, the propagation direction changes according to

$$n_1 \sin \theta_1 = n_2 \sin \theta_2 \quad (1.16)$$

where θ_1 is the angle of incidence and θ_2 is the angle of emergence (see Figure 1.10).

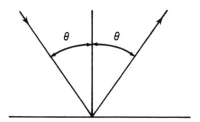

Fig. 1.8.   *The law of reflection*

   Fig. 1.9.   *Scattering from a rough surface*

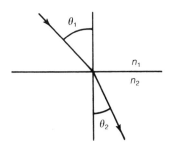

Fig. 1.10.   *The law of refraction*

From Equation (1.16) we see that when $n_1 > n_2$, we can have $\theta_2 = \pi/2$. This occurs for an angle of incidence called the critical angle given by

$$\sin \theta_1 = \frac{n_2}{n_1} \tag{1.17}$$

This is called total internal reflection and will be treated in more detail in Section 9.4.

Finally, we also mention that for light reflected at the interface in Figure 1.10, when $n_1 < n_2$, the phase is changed by π.

1.9   THE SIMPLE CONVEX (POSITIVE) LENS

We shall here not go into the general theory of lenses, but just mention some of the more important properties of a simple, convex, ideal lens. For more details, see Chapter 2 and Section 4.6.

Fig. 1.11

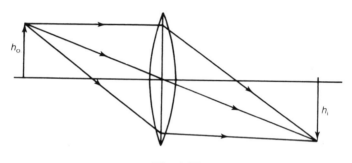

Fig. 1.12

Figure 1.11 illustrates the imaging property of the lens. From an object point P_o, light rays are emitted in all directions. That this point is imaged means that all rays from P_o which pass the lens aperture D intersect at an image point P_i.

To find P_i, it is sufficient to trace just two of these rays. Figure 1.12 shows three of them. The distance b from the lens to the image plane is given by the lens formula

$$\frac{1}{a} + \frac{1}{b} = \frac{1}{f} \tag{1.18}$$

and the transversal magnification

$$m = \frac{h_i}{h_o} = \frac{b}{a} \tag{1.19}$$

In Figure 1.13(a), the case of a point source lying on the optical axis forming a spherical diverging wave that is converted to a converging wave and focuses onto a point on the optical axis is illustrated. In Figure 1.13(b) the point source is lying on-axis at a distance from the lens equal to the focal length f. We then get a plane wave that propagates along the optical axis. In Figure 1.13(c) the point source is displaced along the focal plane a distance h from the optical axis. We then get a plane wave propagating in a direction that

Fig. 1.13

Fig. 1.14.   *A plane wave set-up*

makes an angle θ to the optical axis where

$$\tan \theta = h/f \tag{1.20}$$

1.10   A PLANE-WAVE SET-UP

Finally, we refer to Figure 1.14 which shows a commonly applied set-up to form a uniform, expanded plane wave from a laser beam. The laser beam is a plane wave with a small cross-section, typically 1 mm. To increase the cross-section, the beam is first directed through lens L_1, usually a microscope objective which is a lens of very short focal length f_1. A lens L_2 of greater diameter and longer focal length f_2 is placed as shown in the figure. In the focal point of L_1 a small opening (a pinhole) of diameter typically 10 μm is placed. In that way, light which does not fall at the focal point is blocked. Such stray light is due to dust and impurities crossed by the laser beam on its way via other optical elements (like mirrors, beamsplitters, etc.) and it causes the beam not to be a perfect plane wave.

PROBLEMS

1.1 How many 'yellow' light waves ($\lambda = 550$ nm) will fit into a distance in space equal to the thickness of a piece of paper (0.1 mm)? How far will the same number of microwaves ($v = 10^{10}$ Hz, i.e 10 GHz, and $v = 3 \times 10^8$ m/s) extend?

1.2 Using the wave functions

$$\psi_1 = 4\sin 2\pi(0.2z - 3t)$$

$$\psi_2 = \frac{\sin(7z + 3.5t)}{2.5}$$

determine in each case (a) the frequency, (b) wavelength, (c) period, (d) amplitude, (e) phase velocity and (f) direction of motion. Time is in seconds and z in metres.

1.3 Consider the plane electromagnetic wave (in SI units) given by the expressions $U_x = 0$, $U_y = \exp i[2\pi \times 10^{14}(t - x/c) + \pi/2]$, and $U_z = 0$.
   What is the frequency, wavelength, direction of propagation, amplitude and phase constant of the wave?

1.4 A plane, harmonic light wave has an electric field given by

$$U_z = U_0 \exp i\left[\pi 10^{15}\left(t - \frac{x}{0.65c}\right)\right]$$

while travelling in a piece of glass. Find

(a) the frequency of the light,
(b) its wavelength,
(c) the index of refraction of the glass.

1.5 Imagine that we have a non-absorbing glass plate of index n and thickness Δz which stands between a source and an observer.

(a) If the unobstructed wave (without the plate present) is $U_u = U_0 \exp i\omega(t - z/c)$, ($\omega = 2\pi v$) show that with the plate in place the observer sees a wave

$$U_p = U_0 \exp i\omega\left[t - \frac{(n-1)\Delta z}{c} - \frac{z}{c}\right]$$

(b) Show that if either $n \approx 1$ or Δz is very small, then

$$U_p = U_u + \frac{\omega(n-1)\Delta z}{c} U_u e^{-i\pi/2}$$

The second term on the right may be interpreted as the field arising from the oscillating molecules in the glass plate.

1.6 Show that the optical path, defined as the sum of the products of the various indices times the thicknesses of media traversed by a beam, that is, $\sum_i n_i x_i$, is equivalent to the length of the path in vacuum which would take the same time for that beam to travel.

1.7 Write down an equation describing a sinusoidal plane wave in three dimensions with wavelength λ, velocity v, propagating in the following directions:

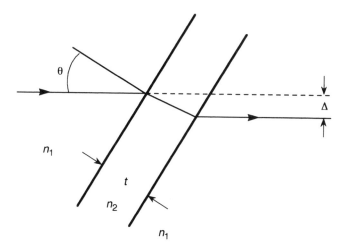

Fig. P1.1

a) $+z$-axis
b) Along the line $x = y$, $z = 0$
c) Perpendicular to the planes $x + y + z = $ const.

1.8 Show that the rays from a point source S that are reflected by a plane mirror appear to be coming from the image point S'. Locate S'.

1.9 Consider Figure P1.1. Calculate the deviation Δ produced by the plane parallel slab as a function of n_1, n_2, t, θ.

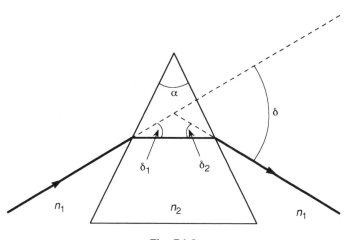

Fig. P1.2

1.10 The deviation angle δ gives the total deviation of a ray incident onto a prism, see Figure P1.2. It is given by $\delta = \delta_1 + \delta_2$. Minimum deviation occurs when $\delta_1 = \delta_2$.

(a) Show that in this case δ_m, the value of δ, obeys the equation

$$\frac{n_2}{n_1} = \frac{\sin\frac{1}{2}(\alpha + \delta_m)}{\sin\frac{1}{2}\alpha}$$

(b) Find δ_m for $\alpha = 60°$ and $n_2/n_1 = 1.69$.

1.11(a) Starting with Snell's law prove that the vector refraction equation has the form

$$n_2\mathbf{k}_2 - n_1\mathbf{k}_1 = (n_2\cos\theta_2 - n_1\cos\theta_1)\mathbf{u}_n$$

where $\mathbf{k}_1, \mathbf{k}_2$ are unit propagation vectors and $\mathbf{u}_n$ is the surface normal pointing from the incident to the transmitting medium.
(b) In the same way, derive a vector expression equivalent to the law of reflection.

2 Gaussian optics

2.1 REFRACTION AT A SPHERICAL SURFACE

Consider Figure 2.1 where we have a sphere of radius R centred at C and with refractive index n'. The sphere is surrounded by a medium of refractive index n. A light ray making an angle α with the z-axis is incident on the sphere at a point A at height x above the z-axis. The ray is incident on a plane which is normal to the radius R and the angle of incidence θ is the angle between the ray and the radius from C. The angle of refraction is θ' and the refracted ray is making an angle α' with the z-axis. By introducing the auxiliary angle ϕ we have the following relations:

$$\phi = \theta' - \alpha' \tag{2.1a}$$

$$\phi = \theta - \alpha \tag{2.1b}$$

$$\sin \phi = \frac{x}{R} \tag{2.1c}$$

$$n \sin \theta = n' \sin \theta' \tag{2.1d}$$

The last equation follows from Snell's law of refraction. By assuming the angles to be small we have $\sin \phi \approx \phi$, $\sin \theta \approx \theta$, $\sin \theta' \approx \theta'$ and by combining Equations (2.1) we get the relation

$$\alpha' = \frac{n - n'}{n'R} x + \frac{n}{n'} \alpha = -\frac{P}{n'} x + \frac{n}{n'} \alpha \tag{2.2}$$

where

$$P = \frac{n' - n}{R} \tag{2.3}$$

is called the power of the surface.

The spherical surface in Figure 2.1 might be the front surface of a spherical lens. In tracing rays through optical systems it is important to maintain consistent sign conventions. It is common to define ray angles as positive counterclockwise from the z-axis and negative in the opposite direction. It is also common to define R as positive when the vertex V of the surface is to the left of the centre C and negative when it is to the right of C.

As can be realized, a ray is completely determined at any plane normal to the z-axis by specifying x, its height above the z-axis in that plane, and its angle α relative to the z-axis. A ray therefore can be specified by a column matrix

$$\begin{pmatrix} x \\ \alpha \end{pmatrix}$$

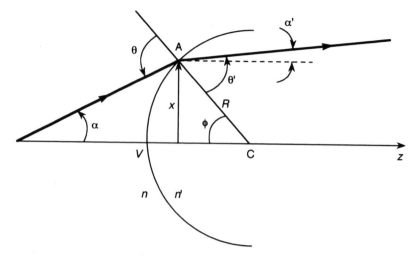

Fig. 2.1. *Refraction at a spherical interface*

The two components of this matrix will be altered as the ray propagates through an optical system. At the point A in Figure 2.1 the height is unaltered, and this fact can be expressed as

$$x' = x \tag{2.4}$$

The transformation at this point can therefore be expressed in matrix form as

$$\begin{pmatrix} x' \\ \alpha' \end{pmatrix} = R \begin{pmatrix} x \\ \alpha \end{pmatrix} \tag{2.5}$$

where

$$R = \begin{pmatrix} 1 & 0 \\ -\dfrac{P}{n'} & \dfrac{n}{n'} \end{pmatrix} \tag{2.6}$$

is the refraction matrix for the surface.

At this point it is appropriate to point out the approximations involved in reaching this formula. First, we have assumed the ray to lie in the xz-plane. To be general we should have considered the ray to lie in an arbitrary plane, taken its components in the xz- and yz-planes and introduced the component angles α and β relative to the z-axis. We then would have found that x and α at a given point depend only on x and α at other points, not on y and β. In other words, the pairs of variables (x, α) and (y, β) are decoupled from one another and may be treated independently. This is true only within the assumption of small angles. Because of this independence it is not necessary to perform calculations on both projections simultaneously. We do the calculations on the projection in the xz-plane and the answers will also apply for the yz-plane with the substitutions $x \rightarrow y$ and $\alpha \rightarrow \beta$. The xz projections behave as though y and β were zero. Such rays, which lie in a single plane containing the z-axis are called meridional rays.

In this theory we have assumed that an optical axis can be defined and that all light rays and all normals to refracting or reflecting surfaces make small angles with the axis. Such

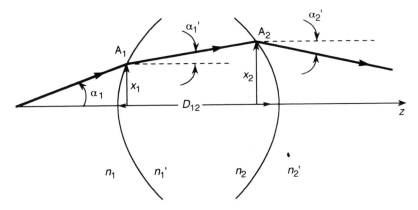

Fig. 2.2. *Ray tracing through a spherical lens*

light rays are called paraxial rays. This first-order approximation was first formulated by C. F. Gauss and is therefore often termed Gaussian optics.

After these remarks we proceed by considering the system in Figure 2.2 consisting of two refracting surfaces with radii of curvature R_1 and R_2 separated by a distance D_{12}. The transformation at the first surface can be written as

$$\begin{pmatrix} x'_1 \\ \alpha'_1 \end{pmatrix} = R_1 \begin{pmatrix} x_1 \\ \alpha_1 \end{pmatrix} \quad \text{with} \quad R_1 = \begin{pmatrix} 1 & 0 \\ -\dfrac{P_1}{n'_1} & \dfrac{n_1}{n'_1} \end{pmatrix} \tag{2.7}$$

where

$$P_1 = \frac{n'_1 - n_1}{R_1} \tag{2.8}$$

The translation from A_1 to A_2 is given by

$$x_2 = x'_1 + D_{12}\alpha'_1 \tag{2.9a}$$

$$\alpha_2 = \alpha'_1 \tag{2.9b}$$

which can be written in matrix form as

$$\begin{pmatrix} x_2 \\ \alpha_2 \end{pmatrix} = T_{12} \begin{pmatrix} x'_1 \\ \alpha'_1 \end{pmatrix} \quad \text{with} \quad T_{12} = \begin{pmatrix} 1 & D_{12} \\ 0 & 1 \end{pmatrix} \tag{2.10}$$

The refraction at A_2 is described by

$$\begin{pmatrix} x'_2 \\ \alpha'_2 \end{pmatrix} = R_2 \begin{pmatrix} x_2 \\ \alpha_2 \end{pmatrix} \quad \text{with} \quad R_2 = \begin{pmatrix} 1 & 0 \\ -\dfrac{P_2}{n'_2} & \dfrac{n_2}{n'_2} \end{pmatrix} \tag{2.11}$$

where

$$P_2 = \frac{n'_2 - n_2}{R_2} \tag{2.12}$$

These equations may be combined to give the overall transformation from a point just to the left of A_1 to a point just to the right of A_2:

$$\begin{pmatrix} x_2' \\ \alpha_2' \end{pmatrix} = M_{12} \begin{pmatrix} x_1 \\ \alpha_1 \end{pmatrix} \qquad \text{with} \qquad M_{12} = R_2 T_{12} R_1 \tag{2.13}$$

This process can be repeated as often as necessary. The linear transformation between the initial position and angle x, α and the final position and angle x', α' can then be written in the matrix form

$$\begin{pmatrix} x' \\ \alpha' \end{pmatrix} = M \begin{pmatrix} x \\ \alpha \end{pmatrix} \tag{2.14}$$

where M is the product of all the refraction and translation matrices written in order, from right to left, in the same sequence followed by the light ray.

The determinant of M is the product of all the determinants of the refraction and translation matrices. We see from Equation (2.10) that the determinant of a translation matrix is always unity and from Equation (2.6) that the determinant of a refraction matrix is given by the ratio of initial to final refractive indices. Thus the determinant of M is the product of the determinants of the separate refraction matrices and takes the form

$$\det M = \left(\frac{n_1}{n_1'}\right)\left(\frac{n_2}{n_2'}\right)\dots \tag{2.15}$$

But $n_1' = n_2$, $n_2' = n_3$ and so on, leaving us with

$$\det M = \frac{n}{n'} \tag{2.16}$$

where n is the index of the medium to the left of the first refracting surface, and n' is the index of the medium to the right of the last refracting surface.

2.1.1 Examples

(1) *Simple lens.* The matrix M is the same as M_{12} in Equation (2.13). By performing the matrix multiplication using $n_1' = n_2$, $n_1 = n$, $n_2' = n'$ and $D_{12} = d$, we get

$$M = \begin{pmatrix} 1 - \dfrac{P_1 d}{n_2} & \dfrac{nd}{n_2} \\[3mm] -\dfrac{P_2}{n'} + \dfrac{P_1 P_2 d}{n' n_2} - \dfrac{P_1}{n'} & \dfrac{n}{n'}\left(1 - \dfrac{P_2 d}{n_2}\right) \end{pmatrix} \tag{2.17}$$

(2) *Thin lens.* A thin lens is a simple lens with a negligible thickness d. If we let $d \to 0$ (i.e $d \ll R$) in Equation (2.17) we obtain

$$M = \begin{pmatrix} 1 & 0 \\[2mm] -\dfrac{P}{n'} & \dfrac{n}{n'} \end{pmatrix} \tag{2.18}$$

where the total power is given by (remember the sign convention for R)

$$P = P_1 + P_2 = \frac{n_2 - n}{R_1} + \frac{n' - n_2}{R_2} \tag{2.19}$$

Note that M has the same form for a thin lens as for a single refracting surface. Note also that the matrix elements $M_{11} = 1$ and $M_{12} = 0$. This means that we have $x' = x$, independently of the value of α.

2.2 THE GENERAL IMAGE-FORMING SYSTEM

In a general image-forming system (possibly consisting of several lens elements) an incoming ray at point B is outgoing from point B', shown schematically in Figure 2.3. The transformation matrix from B to B' is

$$M = \begin{pmatrix} M_{11} & M_{12} \\ M_{21} & M_{22} \end{pmatrix} \tag{2.20}$$

where the only requirement so far is

$$\det M = \frac{n}{n'} \tag{2.21}$$

We now ask if it is possible to find new reference planes instead of B and B' for which the general matrix M will take the form of that for a thin lens. These will turn out to be the so-called principal planes and intersect the axis at H and H' in Figure 2.3. The transformation matrix from the H-plane to the H'-plane can be written in terms of M by adding translations T and T':

$$M_{\mathrm{HH'}} = T'MT = \begin{pmatrix} 1 & D' \\ 0 & 1 \end{pmatrix}\begin{pmatrix} M_{11} & M_{12} \\ M_{21} & M_{22} \end{pmatrix}\begin{pmatrix} 1 & D \\ 0 & 1 \end{pmatrix}$$
$$= \begin{pmatrix} M_{11} + D'M_{21} & M_{11}D + M_{12} + D'(M_{21}D + M_{22}) \\ M_{21} & M_{21}D + M_{22} \end{pmatrix} \tag{2.22}$$

The principal planes are defined as planes of unit magnification. Pairs of points in these planes are images of each other and planes with this property are called conjugate planes. Because of this requirement, the 1, 1 element of $M_{\mathrm{HH'}}$ must be unity and the 1, 2 element

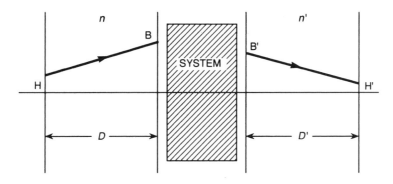

Fig. 2.3

must be zero, giving

$$M_{HH'} = \begin{pmatrix} 1 & 0 \\ M_{21} & \dfrac{n}{n'} \end{pmatrix} \tag{2.23}$$

We now equate the elements of the matrices in Equation (2.22) and (2.23)

$$11: \ M_{11} + D'M_{21} = 1 \quad \text{i.e} \quad D' = \frac{1 - M_{11}}{M_{21}} \tag{2.24a}$$

$$22: \ M_{21}D + M_{22} = \frac{n}{n'} \quad \text{i.e.} \quad D = \frac{(n/n') - M_{22}}{M_{21}} \tag{2.24b}$$

These equations are meaningful only if the condition

$$M_{21} \neq 0 \tag{2.25}$$

is satisfied. This then becomes the requirement that our general Gaussian system be image-forming. (Identification of matrix element 12 gives the same condition). To complete the final equivalence between our general image-forming system and a thin lens, it is only necessary to make the identification

$$-\frac{P}{n'} = M_{21} \tag{2.26}$$

Thus the image-formation condition, Equation (2.25) guarantees that our system has non-zero power. This means that all image forming systems have the same formal behaviour in Gaussian optics, as far as ray-tracing is concerned. It should be noted that for an afocal system like the plane wave set-up in Figure 1.14 where the two focal points coincide, $M_{21} = 0$. This is the same configuration as in a telescope where we only have angular magnification.

2.3   THE IMAGE-FORMATION PROCESS

We now want to move from the principal planes to other conjugate planes and determine the object-image relationships that result. This is done by translation transformations over the distances a and b in Figure 2.4. The overall transformation matrix from A to A' is given by

$$M_{AA'} = \begin{pmatrix} 1 & b \\ 0 & 1 \end{pmatrix} \begin{pmatrix} 1 & 0 \\ -\dfrac{P}{n'} & \dfrac{n}{n'} \end{pmatrix} \begin{pmatrix} 1 & a \\ 0 & 1 \end{pmatrix}$$

$$= \begin{pmatrix} 1 - \dfrac{bP}{n'} & a - \dfrac{abP}{n'} + \dfrac{nb}{n'} \\ -\dfrac{P}{n'} & -\dfrac{aP}{n'} + \dfrac{n}{n'} \end{pmatrix} \tag{2.27}$$

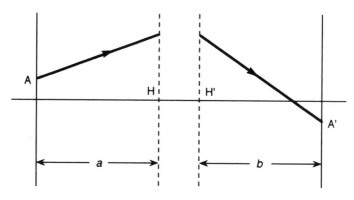

Fig. 2.4

The image-formation condition is that the 1,2 element of this matrix be zero:

$$a - \frac{abP}{n'} + \frac{nb}{n'} = 0 \tag{2.28}$$

that is

$$\frac{n}{a} + \frac{n'}{b} = P \tag{2.29}$$

When the image is at $+\infty$, the object is in the first focal plane at a distance

$$a = \frac{n}{P} \equiv f \tag{2.30}$$

to the left of the first principal plane. When the object is at $+\infty$, the image is in the second focal plane at a distance

$$b = \frac{n'}{P} \equiv f' \tag{2.31}$$

to the right of the second principal plane. Thus Equation (2.29) may be written in the Gaussian form

$$\frac{n}{a} + \frac{n'}{b} = \frac{n}{f} = \frac{n'}{f'} \tag{2.32}$$

When the refractive indices in image and object space are the same ($n = n'$), this equation takes on the well known form

$$\frac{1}{a} + \frac{1}{b} = \frac{1}{f} \tag{2.33}$$

i.e. the lens formula.

When we have image formation, our matrix can be written

$$M_{AA'} = \begin{pmatrix} m_x & 0 \\ -\dfrac{P}{n'} & m_\alpha \end{pmatrix} \tag{2.34}$$

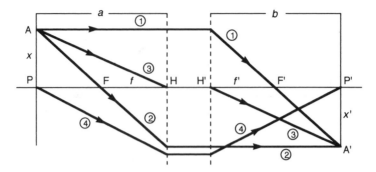

Fig. 2.5.   *Principal planes with some key rays*

where the lateral magnification is

$$m_x = 1 - \frac{bP}{n'} = 1 - \frac{b}{f'} = -\frac{nb}{n'a} \tag{2.35}$$

and the ray angle magnification is

$$m_\alpha = -\frac{aP}{n'} + \frac{n}{n'} = -\frac{a}{b} \tag{2.36}$$

From the condition det $M_{AA'} = n/n'$ we obtain the result

$$m_x m_\alpha = \frac{n}{n'} \tag{2.37}$$

It should be emphasized that the physical location of the principal planes could be inside one of the components of the image-forming system. Or they could be outside. The point to be made is that these are mathematical planes, and the rays behave as though they were deviated as shown in Figure 2.5. There is no *a priori* reason for the order of the principal planes. The plane H could be to the right of H'. The plane H will be to the right of F and H' to the left of F' if f and f' are positive.

2.4   STOPS AND APERTURES

Stops and apertures play an important role in lens systems.

The aperture stop is defined to be the aperture which physically limits the solid angle of rays passing through the system from an on-axis object point. A simple example is shown in Figure 2.6(a) where the hole in the screen limits the solid angle of rays from the object at P_o. The rays are cut off at A and B. The images of A and B are A' and B'. To an observer looking back through the lens from a position near P'_o it will appear as if A' and B' are cutting off the rays. If we move the screen to the left of F, we have the situation shown in Figure 2.6(b). The screen is still the aperture stop, but the images A', B' of A and B are now to the right of P'_o. To an observer who moves sufficiently far to the right it still appears as if the rays are being cut off by A' and B'.

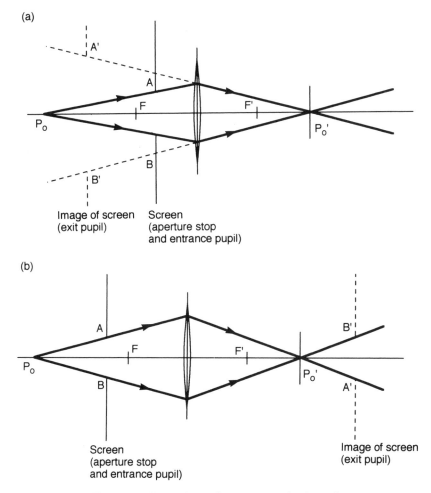

(a)

Image of screen
(exit pupil)

Screen
(aperture stop
and entrance pupil)

(b)

Screen
(aperture stop
and entrance pupil)

Image of screen
(exit pupil)

Fig. 2.6.   *Illustrations of entrance and exit pupils*

A 'space' may be defined that contains all physical objects to the right of the lens plus all points conjugate to physical objects that are to the left of the lens. It is called the image space. In Figures 2.6(a, b) all primed points are in image space. The image of the aperture stop in image space is called the exit pupil. To an observer in image space it appears either as if the rays converging to an on-axis image P_o' are limited in solid angle by the exit pupil $A'B'$ as in Figure 2.6(a) or as if the rays diverging from P_o' are limited in solid angle by $A'B'$ as in Figure 2.6(b).

By analogy to the image space, a space called the object space may be defined that contains all physical objects to the left of the lens plus all points conjugate to any physical object that may be to the right of the lens. In Figure 2.6(a, b) all unprimed objects are in the object space. The image of the aperture stop in the object space is defined as the entrance pupil. The aperture stop in Figure 2.6(a, b) is already in the object space, hence it is itself the entrance pupil.

In a multilens system some physical objects will be neither in the object nor in the image space but in between the elements. If a given point is imaged by all lens elements to its right, it will give an image in the image space; if imaged by all elements to its left, it will give an image in the object space. A systematic method of finding the entrance pupil is to image all stops and lens rims to the left through all intervening refracting elements of the system into the object space and find the solid angle subtended by each at P_o. The one with the smallest solid angle is the entrance pupil, and the physical object corresponding to it is the aperture stop. Alternatively we may image all stops and lens rims to the right through all intervening refractive elements into the image space and determine the solid angle subtended by each image at P'_o. The one with the smallest solid angle is the exit pupil, and the corresponding real physical object is the aperture stop.

2.5 LENS ABERRATIONS

The ray-tracing equations used in the theory of Gaussian optics are correct to first order in the inclination angles of the rays and the normals to refracting or reflecting surfaces. When higher- order approximations are used for the trigonometric functions of the angles, departures from the predictions of Gaussian optics will be found. No longer will it be generally true that all the rays leaving a point object will exactly meet to form a point image or that the magnification in a given transverse plane is constant. Such deviations from ideal Gaussian behaviour are known as lens aberrations. In addition, the properties of a lens system may be wavelength- dependent, known as chromatic aberrations.

Monochromatic aberrations may be treated mathematically in lowest order by carrying out the ray-tracing calculations to third order in the angles. The resulting 'third-order theory' is itself valid only for small angles and for many real systems calculations must be carried out to still higher order, say fifth or seventh. (For a centred system with rotational symmetry, only odd powers of the angles will appear in the ray-tracing formulas.)

Most compound lens systems contain enough degrees of freedom in their design to compensate for aberrations predicted by the third-order theory. For real systems the residual higher-order aberrations would still be present, and there are not enough design parameters to eliminate all of them as well. The performance of a lens system must be judged according to the intended use. The criteria for a telescope objective and for a camera lens for close-ups are quite different.

Here we will not treat lens aberrations in any detail. Figure 2.7 illustrates spherical aberration and in Section 10.3.1 distortion is treated in some detail. The other third-order aberrations are called coma, astigmatism and field curvature. Because of the complexity of the higher-order aberrations they are usually treated numerically. Now lens design computer programs are available commercially. Such programs trace a lot of different rays through the system and the points where they intersect the image plane is called a spot diagram. By changing the design parameters, the change in the spot diagram can be observed. Some computer programs do such analyses automatically. The computer is given a quality factor (or merit function) of some sort, which means how much of each aberration is tolerated. Then a roughly designed system which, in the first approximation, meets the particular requirements is given as input. The computer will then trace several rays through the system and evaluate the image errors. After perhaps twenty or more iterations, it will have changed the initial configuration so that it now meets the specified

Fig. 2.7. *Spherical aberration. The focus of the paraxial rays is at P_o'. The marginal rays focus at a point closer to the lens*

limits on aberrations. However, a quality factor is somewhat like a crater-pocket surface in a multidimensional space. The computer will carry the design from one hole to the next until it finds one deep enough to meet the specifications. There is no way to tell if that solution corresponds to the deepest hole without sending the computer out again and again meandering along totally different routes.

PROBLEMS

2.1 Verify directly by matrix methods that use of the matrix in Equation (2.34) will yield values of (x', α') for rays 1, 2, 3, 4 in Figure 2.5 so that they behave as shown.

2.2 Consider the system shown in Figure P2.1 where the focal lengths of the first system are f_1, f_1' and those of the second f_2, f_2'. The respective powers are

$$P_1 = \frac{n_1}{f_1} = \frac{n_1'}{f_1'}$$

$$P_2 = \frac{n_2}{f_2} = \frac{n_2'}{f_2'}$$

(a) Find the transformation matrix $M_{H_1 H_2'}$ between the first principal plane of the first system and the second principal plane of the second system.

We denote the principal planes of the whole system by H and H' and the distances $HH_1 = D$ and $H_2' H' = D'$.

Fig. P2.1

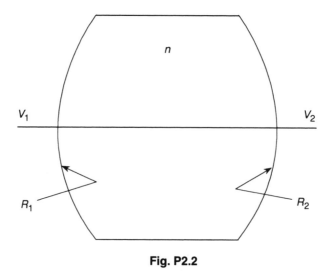

Fig. P2.2

(b) Express the transformation matrix $M_{HH'}$ between H and H' in terms of the total power P and n and m'

(c) Find the total power of the system

(d) Find D and D'.

2.3 A doublet consists of two lenses with principal plane separation $d = f'_1 + f_2 + l$, see Figure P2.1. We set $n_2 = n'_2 = 1$.

(a) Find the power P of the doublet in terms of P_1, P_2 and l.

(b) Find the first and second focal lengths.

2.4 Find the power and the locations of the principal planes for a combination of two thin lenses each with the same focal length $f > 0$ separated by a distance d: (a) where $d = f$, (b) where $d = 3f/4$.

2.5 Show that the combination of two lenses having equal and opposite powers a finite, positive, distance d apart has a net positive power P, and find P as a function of d.

2.6 A thick lens as shown in Figure P2.2, is used in air. The first and second radii of curvature are $R_1 > 0$ and $R_2 < 0$, the index is $n > 1$, and the thickness $|V_1 V_2|$ is d. What will be the aperture stop for this lens for an axial object at a general distance S_1 to the left of V_1? Is the aperture stop always the same? (No calculation is necessary to solve this problem.)

2.7 A thin lens L_1 with a 5.0 cm diameter aperture and focal length $+4.0$ cm is placed 4.0 cm to the left of another lens L_2 4.0 cm in diameter with a focal length of $+10.0$ cm. A 2.0 cm high object is located with its centre on the axis 5 cm in front of L_1. There is a 3.0 cm diameter stop centered halfway between L_1 and L_2. Find the position and size of (a) the entrance pupil, (b) the exit pupil, (c) the image. Make a brief sketch to scale.

3 Interference

3.1 GENERAL DESCRIPTION

Interference can occur when two or more waves overlap each other in space.
  Assume that two waves described by

$$u_1 = U_1 e^{i\phi_1} \tag{3.1a}$$

and

$$u_2 = U_2 e^{i\phi_2} \tag{3.1b}$$

overlap.
  The electromagnetic wave theory tells us that the resulting field simply becomes the
sum, viz.

$$u = u_1 + u_2 \tag{3.2}$$

The observable quantity is, however, the intensity, which becomes

$$I = |u|^2 = |u_1 + u_2|^2 = U_1^2 = U_1^2 + U_2^2 + 2U_1 U_2 \cos(\phi_1 - \phi_2)$$
$$= I_1 + I_2 + 2\sqrt{I_1 I_2} \cos \Delta\phi \tag{3.3}$$

where

$$\Delta\phi = \phi_1 - \phi_2 \tag{3.4}$$

As can be seen, the resulting intensity does not become merely the sum of the intensities
$(= I_1 + I_2)$ of the two partial waves. One says that the two waves interfere and
$2\sqrt{I_1 I_2} \cos \Delta\phi$ is called the interference term. We also see that when

$$\Delta\phi = (2n + 1)\pi, \quad \text{for} \quad n = 0, 1, 2, \ldots$$

$\cos \Delta\phi = -1$ and I reaches its minima. The two waves are in antiphase which means that
they interfere destructively.
  When

$$\Delta\phi = 2n\pi, \quad \text{for} \quad n = 0, 1, 2, \ldots$$

$\cos \Delta\phi = 1$ and the intensity reaches its maxima. The two waves are in phase which means
that they interfere constructively.
  For two waves of equal intensity, i.e. $I_1 = I_2 = I_0$, Equation (3.3) becomes

$$I = 2I_0[1 + \cos \Delta\phi] = 4I_0 \cos^2\left(\frac{\Delta\phi}{2}\right) \tag{3.5}$$

where the intensity varies between 0 and $4I_0$.

3.2 COHERENCE

Detection of light (i.e. intensity measurement) is an averaging process in space and time. In developing Equation (3.3) we did no averaging because we tacitly assumed the phase difference $\Delta\phi$ to be constant in time. That means that we assumed u_1 and u_2 to have the same single frequency. Ideally, a light wave with a single frequency must have an infinite length. Mathematically, even a pure sinusoidal wave of finite length will have a frequency spread according to the Fourier theorem (see Appendix B). Therefore, sources emitting light of a single frequency do not exist.

One way of illustrating the light emitted by real sources is to picture it as sinusoidal wave trains of finite length with randomly distributed phase differences between the individual trains.

Assume that we apply such a source in an interference experiment, e.g. the Michelson interferometer described in Section 3.5.2. Here the light is divided into two partial waves of equal amplitudes by a beamsplitter whereafter the two waves are recombined to interfere after having travelled different paths.

In Figure 3.1 we have sketched two successive wave trains of the partial waves. The two wave trains have equal amplitude and length L_c, with an abrupt, arbitrary phase difference. Figure 3.1(a) shows the situation when the two partial waves have travelled equal path lengths. We see that although the phase of the original wave fluctuates randomly, the phase difference between the partial waves 1 and 2 remains constant in time. The resulting intensity is therefore given by Equation (3.3). Figure 3.1(c) shows the situation when partial wave 2 has travelled a path length L_c longer than partial wave 1. The head of the wave trains in partial wave 2 then coincide with the tail of the corresponding wave trains in partial wave 1. The resulting instantaneous intensity is still

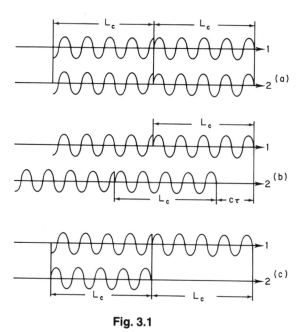

Fig. 3.1

given by Equation (3.3), but now the phase difference fluctuates randomly as the successive wave trains pass by. As a result, $\cos\Delta\phi$ varies randomly between $+1$ and -1. When averaged over many wave trains, $\cos\Delta\phi$ therefore becomes zero and the resulting, observable intensity will be

$$I = I_1 + I_2 \tag{3.6}$$

Figure 3.1(b) shows an intermediate case where partial wave 2 has travelled a path length l longer than partial wave 1, where $0 < l < L_c$. Averaged over many wave trains, the phase difference now varies randomly in a time period proportional to $\tau = l/c$ and remains constant in a time period proportional to $\tau_c - \tau$ where $\tau_c = L_c/c$. The result is that we still can observe an interference pattern according to Equation (3.3), but with a reduced contrast. To account for this loss of contrast, Equation (3.3) can be written as

$$I = I_1 + I_2 + 2\sqrt{I_1 I_2}\,|\gamma(\tau)|\cos\Delta\phi \tag{3.7}$$

where $|\gamma(\tau)|$ means the absolute value of $\gamma(\tau)$.

To see clearly that this quantity is related to the contrast of the pattern, we introduce the definition of contrast or visibility

$$V = \frac{I_{\max} - I_{\min}}{I_{\max} + I_{\min}} \tag{3.8}$$

where $I_{\max}$ and $I_{\min}$ are two neighbouring maxima and minima of the interference pattern described by Equation (3.7). Since $\cos\Delta\phi$ varies between $+1$ and -1 we have

$$I_{\max} = I_1 + I_2 + 2\sqrt{I_1 I_2}\,|\gamma(\tau)| \tag{3.9a}$$

$$I_{\min} = I_1 + I_2 + 2\sqrt{I_1 I_2}\,|\gamma(\tau)| \tag{3.9b}$$

which, put into Equation (3.8), gives

$$V = \frac{2\sqrt{I_1 I_2}\,|\gamma(\tau)|}{I_1 + I_2} \tag{3.10}$$

For two waves of equal intensity, $I_1 = I_2$, and Equation (3.10) becomes

$$V = |\gamma(\tau)| \tag{3.11}$$

which shows that in this case $|\gamma(\tau)|$ is exactly equal to the visibility. $\gamma(\tau)$ is termed the complex degree of coherence and is a measure of the ability of the two wave fields to interfere. From the previous discussions we must have

$$|\gamma(0)| = 1 \tag{3.12a}$$

$$|\gamma(\tau_c)| = 0 \tag{3.12b}$$

$$0 \leqslant |\gamma(\tau)| \leqslant 1 \tag{3.12c}$$

where Equations (3.12a) and (3.12b) represent the two limiting cases of complete coherence and incoherence respectively, while inequality (3.12c) represents partial coherence.

Of more interest is to know the value of τ_c, i.e. at which path length difference $|\gamma(\tau)| = 0$. In Section 5.3.9 we find that in the case of a two-frequency laser this happens when

$$\tau = \tau_c = \frac{L_c}{c} = \frac{1}{\Delta v} \tag{3.13}$$

where Δv is the difference between the two frequencies. It can be shown that this relation applies to any light source with a frequency distribution of width Δv. L_c is termed the coherence length and τ_c the coherence time.

We see that Equation (3.13) is in accordance with our previous discussion where we argued that sources of finite spectral width will emit wave trains of finite length. This is verified by the relation

$$\Delta v = \frac{\Delta \lambda_c}{\lambda^2} \tag{3.14}$$

which can be derived from Equation (1.2).

As given in Section 1.1, the visible spectrum ranges from 4.3 to 7.5×10^{14} Hz which gives a spectral width roughly equal to $\Delta v = 3 \times 10^{14}$ Hz. From Equation (3.13), the coherence time of white light is therefore about 3×10^{-15} s, which corresponds to a coherence length of about 1 µm. In white-light interferometry it is therefore difficult to observe more than two or three interference fringes. This condition can be improved by applying colour filters at the cost of decreasing the intensity.

Ordinary discharge lamps have spectral widths corresponding to coherence lengths of the order of 1 µm while the spectral lines emitted by low-pressure isotope lamps have coherence lengths of several millimetres.

By far the most coherent light source is the laser. A single-frequency laser can have coherence lengths of several hundred metres. This will be analysed in more detail in Section 5.3.9.

So far we have been discussing the coherence between two wave fields at one point in space. This phenomenon is termed temporal or longitudinal coherence. It is also possible to measure the coherence of a wave field at two points in space. This phenomenon is called spatial or transverse coherence and can be analysed by the classical Young's double slit (or pinhole) experiment (see Section 3.5.1). Here the wave field at two points P_1 and P_2 is analysed by passing the light through two small holes in a screen S_1 at P_1 and P_2 and observing the resulting interference pattern on a screen S_2 (see Figure 3.12(a)). In the same way as the temporal degree of coherence $\gamma(\tau)$ is a measure of the fringe contrast as a function of time difference τ, the spatial degree of coherence γ_{12} is a measure of the fringe contrast of the pattern on screen S_2 as a function of the spatial difference D between P_1 and P_2. Note that since γ_{12} is the spatial degree of coherence for $\tau = 0$, it is the contrast of the central fringe on S_2 that has to be measured.

To measure the spatial coherence of the source itself, screen S_1 has to be placed in contact with the source. It is immediately clear that for an extended thermal light source, $|\gamma_{12}| = 0$ unless $P_1 = P_2$, which gives $|\gamma_{11}| = 1$. On the other hand, if we move S_1 away from this source, we observe that $|\gamma_{12}|$ might be different from zero, which shows that a wave field increases its spatial coherence by mere propagation. We also observe that $|\gamma_{12}|$ increases by stopping down the source by, for example, an aperture until $|\gamma_{12}| = 1$ for a pinhole aperture. The distance D_c between P_1 and P_2 for which $|\gamma_{12}| = 0$ is called the spatial coherence length. It can be shown that D_c is inversely proportional to the diameter of the aperture in analogy with the temporal coherence length, which is inversely proportional to the spectral width. Moreover, it can be shown that $|\gamma_{12}|$ is the Fourier transform of the intensity distribution of the source and that $|\gamma(\tau)|$ is the Fourier transform of the spectral distribution of the source (see Section 3.6.)

An experimentalist using techniques like holography, moiré, speckle and photoelasticity need not worry very much about the details of coherence theory. Both in theory and experiments one usually assumes that the degree of coherence is either one or zero. However, one should be familiar with fundamental facts such as:

(1) Light from two separate sources does not interfere.

(2) The spatial and temporal coherence of light from an extended thermal source is increased by stopping it down and by using a colour filter respectively.

(3) The visibility function of a multimode laser exhibits maxima at an integral multiple of twice the cavity length (see Section 5.3.9).

3.3 INTERFERENCE BETWEEN TWO PLANE WAVES

Figure 3.2(a) shows two plane waves u_1, u_2 with propagation directions $\mathbf{n}_1, \mathbf{n}_2$ that lie in the xz-plane making the angles θ_1 and θ_2 to the z-axis. We introduce the following quantities (see Figure 3.2(b)): $\alpha =$ the angle between $\mathbf{n}_1$ and $\mathbf{n}_2$, $\theta =$ the angle between the line bisecting α and the z-axis. The complex amplitude of the two plane waves then

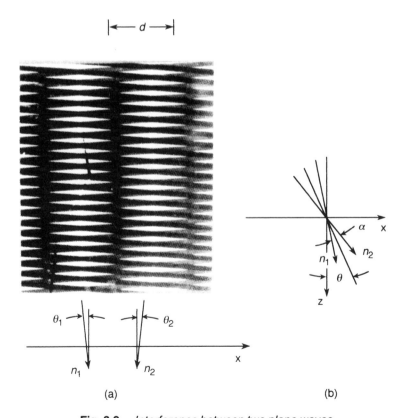

(a)   (b)

Fig. 3.2. *Interference between two plane waves*

becomes (see Equation (1.9a))

$$u_1 = U_1 e^{i\phi_1} \tag{3.15}$$

$$u_2 = U_2 e^{i\phi_2} \tag{3.16}$$

where

$$\phi_1 = k\left[x\sin\left(\theta - \frac{\alpha}{2}\right) + z\cos\left(\theta - \frac{\alpha}{2}\right) \right] \tag{3.17}$$

$$\phi_2 = k\left[x\sin\left(\theta + \frac{\alpha}{2}\right) + z\cos\left(\theta + \frac{\alpha}{2}\right) \right] \tag{3.18}$$

The intensity is given by the general expression in Equation (3.3) by inserting

$$\Delta\phi = \phi_1 - \phi_2 = k\left\{ x\left[\sin\left(\theta - \frac{\alpha}{2}\right) - \sin\left(\theta + \frac{\alpha}{2}\right) \right] + z\left[\cos\left(\theta - \frac{\alpha}{2}\right) - \cos\left(\theta + \frac{\alpha}{2}\right) \right] \right\}$$

$$= 2k\sin\frac{\alpha}{2}\{ - x\cos\theta + z\sin\theta\} \tag{3.19}$$

The interference term is therefore of the form

$$\cos\frac{2\pi}{d}(z\sin\theta - x\cos\theta) \tag{3.20}$$

By comparing this expression with the real part of Equation (1.9a), we see that Equation (3.20) can be regarded as representing a plane wave with its propagation direction lying in the xz-plane making an angle θ with the x-axis as depicted in Figure 3.3, and with a wavelength equal to

$$d = \frac{\lambda}{2\sin(\alpha/2)} \tag{3.21}$$

This is also clearly seen from Figure 3.2. From Equation (3.21) we see that the distance between the interference fringes (the wavelength d) is dependent only on the angle between n_1 and n_2. By comparing Figures 3.2 and 3.4 we see how d decreases as α increases. The diagram in Figure 3.5 shows the relation between d and α and $f = 1/d$ and α according to Equation (3.21). Here we have put $\lambda = 0.6328\,\mu m$, the wavelength of the He–Ne laser.

Fig. 3.3

Fig. 3.4

The intensity distribution across the xy-plane is found by inserting $z = 0$ into Equation (3.19):

$$I = I_1 + I_2 + 2\sqrt{I_1 I_2}\cos\left[2kx\sin\frac{\alpha}{2}\cos\theta\right]\tag{3.22}$$

From the maxima (or minima) of this equation, we find the inter-fringe distance measured along the x-axis to be

$$d_x = \frac{1}{\sin\theta_2 - \sin\theta_1} = \frac{\lambda}{2\sin\frac{\alpha}{2}\cos\theta} = d/\cos\theta\tag{3.23}$$

The second equality of this expression is found by trigonometric manipulation of the angles (see Figure 3.2(b)). Accordingly, the spatial frequency becomes

$$f_x = 1/d_x = \frac{2\sin\frac{\alpha}{2}\cos\theta}{\lambda} = \cos\theta/d\tag{3.24}$$

For completeness, we also quote the definition of the instantaneous frequency of a sinusoidal grating with phase $\phi(x)$ at a point x_0.

$$f_x(x = x_0) = \left.\frac{\mathrm{d}\phi(x)}{\mathrm{d}x}\right|_{x = x_0}\tag{3.25}$$

The intensity distribution given in Equation (3.22) is sketched in Figure 3.6. We see that it varies between

$$I_{max} = I_1 + I_2 + 2\sqrt{I_1 I_2}\tag{3.26}$$

and

$$I_{min} = I_1 + I_2 - 2\sqrt{I_1 I_2}\tag{3.27}$$

with a mean value equal to

$$I_0 = I_1 + I_2\tag{3.28}$$

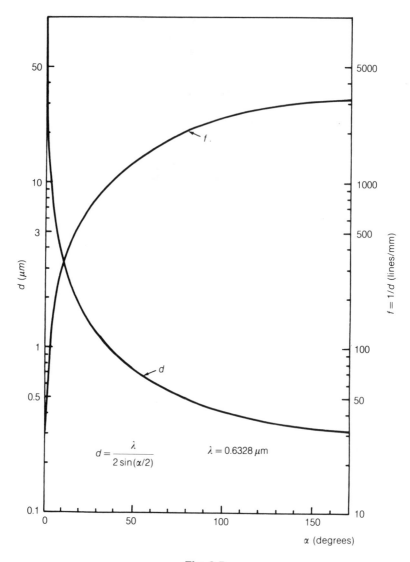

Fig. 3.5

When Equations (3.26) and (3.27) are put into the expression for the visibility or contrast defined in Section 3.2, Equation (3.8), they give

$$V = \frac{I_{max} - I_{min}}{I_{max} + I_{min}} = \frac{2\sqrt{I_1 I_2}}{I_1 + I_2} \tag{3.29}$$

V is equal to the amplitude of the distribution divided by the mean value and varies between 0 and 1. We see that

$$V = 1 \quad \text{for} \quad I_1 = I_2$$
$$V = 0 \quad \text{for} \quad \text{either } I_1 \text{ or } I_2 = 0$$

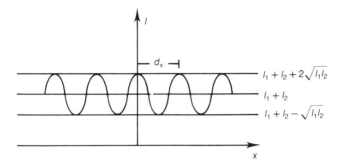

Fig. 3.6. *Intensity distribution in the xy-plane from interference between two plane waves*

3.4 *INTERFERENCE BETWEEN OTHER WAVES*

Figure 3.7 shows the geometric configuration of the fringe pattern in the xz-plane when two spherical waves from two point sources P_1 and P_2 on the z-axis interfere. From Figures 3.7(a) and 3.7(b) we see how the density of the fringes increases as the distance between P_1 and P_2 increases. Note that the figure shows the situation for the actual wavelength. The distance between the point sources in Figure 3.7(a) is about seven wavelengths, which for light with $\lambda = 0.5\,\mu$m would give 3.5 μm. Two real point sources separated by the same distance as in the figure therefore would have resulted in a pattern of much higher density; but the form of pattern would be the same.

Figure 3.8 shows the interference pattern in the xz-plane when a spherical wave from a point P on the z-axis interferes with a plane wave propagating in the z-direction.

In the same way as in the case of two plane waves, we can observe the intensity distribution over a plane of arbitrary orientation in space. A special distribution can be observed over the xy-plane in Figure 3.8. This is further illustrated in Figure 3.9 which shows the case of a spherical wave and a plane wave. The intensity distribution in the xy-plane is given as

$$I = I_1 + I_2 + 2\sqrt{I_1 I_2}\cos(\beta r^2) \tag{3.30}$$

with $r^2 = x^2 + y^2, \beta = $ constant. This is a sinusoidal pattern of linearly increasing frequency and is called a circular zone plate pattern.

So far we have been dealing with interference between combinations of spherical waves and plane waves. An important point to note is that, by measuring the distance between interference fringes over selected planes in space, quantities such as the angle between the propagation directions of plane waves, the distance between point sources and the distance from a point source to the plane of observation can be determined. One further step would be to apply the same type of experiment to more complicated waveforms, such as a wave reflected from a rough surface as indicated in Figure 1.5(c). By observing the interference between the reflected wave and a plane wave, one should in principle be able to determine the topography of the surface. However, for surfaces of roughness greater than the wavelength, phenomena such as interference between light scattered from

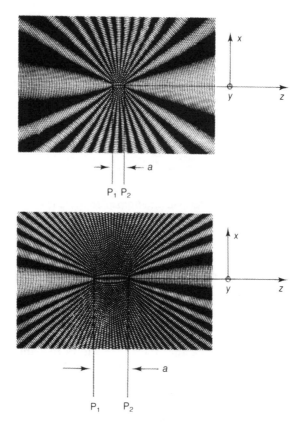

Fig. 3.7.   *Interference between two spherical waves emitted from P_1 and P_2*

Fig. 3.8.   *Interference between a plane wave and a spherical wave*

different points on the surface, multiple scattering and diffraction effects will occur. In this case, it therefore becomes impossible to derive the surface topography from a given interference pattern. For smoother surfaces, however, such as optical components (lenses, mirrors, etc.) where tolerances of the order of fractions of a wavelength are to be measured, that kind of interferometry is quite common.

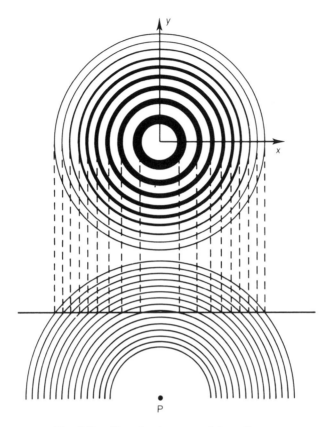

Fig. 3.9.   *The circular zone plate pattern*

Fig. 3.10

Figure 3.10 shows a Michelson interferometer (see Section 3.5.2, Figure 3.14) where the mirrors are exchanged for two non-optical surfaces A_1 and A_2. If these surfaces are identical, it should be possible to observe interference between the waves scattered from A_1 and A_2 regardless of the complexity of the scattered wavefronts. In the case of a plastic deformation of, for example, surface A_2, it should be possible to do interferometric measurement of the resulting surface height difference between A_1 and A_2. The problem is, however, that the phrasing 'identical surfaces' in this context must be taken literally, i.e. the microstructure of the two surfaces must be identical. This has to do with the mutual spatial coherence of the two scattered waves. This requirement on the two surfaces makes this interference experiment impracticable. However, when we learn later on about holography, we shall see that this type of measurement becomes more than an imaginary experiment.

3.5 INTERFEROMETRY

Interference phenomena can be observed in interferometers. As stated in Section 3.2, light waves can interfere only if they are emitted by the same source. Most interferometers therefore consist of the following elements, shown schematically in Figure 3.11.

- light source;
- element for splitting the light into two (or more) partial waves;
- different propagation paths where the partial waves undergo different phase contributions;
- element for superposing the partial waves;
- detector for observation of the interference.

Depending on how the light is split, interferometers are commonly classified as wavefront-dividing or amplitude-dividing interferometers; but there are configurations which fall outside this classification.

3.5.1 Wavefront division

As an example of a wavefront-dividing interferometer, consider the oldest of all interference experiments due to Thomas Young (1801) (Figure 3.12). The incident wavefront is divided by passing through two small holes at P_1 and P_2 in a screen S_1. The emerging spherical wavefronts from P_1 and P_2 will interfere, and the resulting interference pattern is

Fig. 3.11

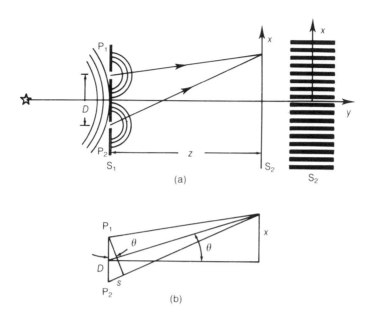

Fig. 3.12. *Young's interferometer*

observed on the screen S_2. This is in analogy with the case of two point sources in Figure 3.7 with the plane of observation oriented parallel to the xy-plane.

The geometric path length difference s of the light reaching an arbitrary point x on S_2 from P_1 and P_2 is found from Figure 3.12(b). When the distance z between S_1 and S_2 is much greater than the distance D between P_1 and P_2, we have, to a good approximation,

$$\frac{s}{D} = \frac{x}{z}, \quad \text{that is} \quad s = \frac{D}{z}x$$

The phase difference therefore becomes

$$\Delta\phi = \frac{2\pi}{\lambda}s = \frac{2\pi D}{\lambda z}x \tag{3.31}$$

which, inserted into the general expression for the resulting intensity distribution, Equation (3.3), gives

$$I(x) = 2I\left(1 + \cos\left(2\pi\frac{D}{\lambda z}x\right)\right) \tag{3.32}$$

We get interference fringes parallel to the y-axis with a spatial period $\lambda z/D$ which decreases as the distance between P_1 and P_2 increases.

Here we have assumed that the waves from P_1 and P_2 are fully coherent. As stated in Section 3.2, this is an ideal case and becomes more and more difficult to fulfil as the distance D between P_1 and P_2 is increased. The contrast of the interference fringes on S_2 is a measure of the degree of coherence. As will be shown in Section 3.6 there is a Fourier transform relationship between the degree of coherence and the intensity distribution of the light source. By assuming the source to be an incoherent circular disc of uniform intensity, one can find the diameter of the source by increasing the separation between P_1 and P_2 until the contrast of the central interference fringe on S_2 vanishes. This is utilized in Michelson's stellar interferometer (Figure 3.13(c)) to measure the diameter of distant stars. This is an extension of Young's interferometer, where a mirror arrangement is used to make the effective distance D sufficiently long.

Other types of wavefront-dividing interferometers are shown in Figure 3.13.

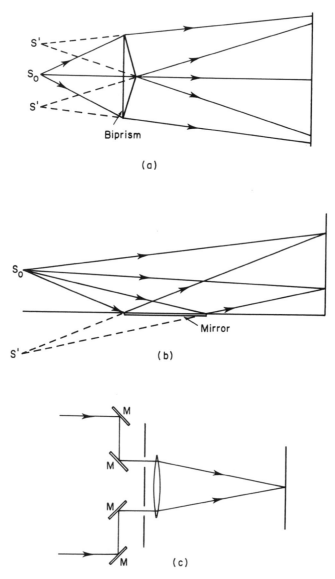

Fig. 3.13. *Examples of wavefront-dividing interferometers: (a) Fresnel biprism; (b) Lloyd's mirror; (c) Michelson's stellar interferometer*

Fig. 3.14. *Michelson's interferometer*

3.5.2 Amplitude division

The most well-known amplitude-dividing interferometer is the Michelson interferometer sketched in Figure 3.14. Here the amplitude of the incident light field is divided by the beamsplitter BS which is partly reflecting. The reflected and the transmitted partial waves propagate to the mirrors M_1 and M_2 respectively, from where they are reflected back and recombine to form the interference distribution on the detector D.

The path-length difference between the two partial waves can be varied by moving one of the mirrors, e.g. M_2, which might be mounted on a movable object. A displacement x of M_2 gives a path length difference $2x$ and a phase difference equal to $\Delta\phi = (2\pi/\lambda)2x$. This results in an intensity distribution given by

$$I(x) = 2I\left(1 + \cos\frac{4\pi x}{\lambda}\right) \tag{3.33}$$

As M_2 moves, its displacement is measured by counting the number of light maxima registered by the detector. By counting the numbers of maxima per unit time, one can find the speed of the object.

Other types of amplitude-dividing interferometers are shown in Figure 3.15.

3.5.3 The dual-frequency Michelson interferometer

In Section 3.2 we stated that two waves of different frequencies do not produce observable interference. By combining two plane waves

$$\psi_1 = e^{i2\pi[(z/\lambda_1) - v_1 t]} \tag{3.34a}$$

and

$$\psi_2 = e^{i2\pi[(z/\lambda_2) - v_2 t]} \tag{3.34b}$$

of different frequencies, the resulting intensity becomes

$$I = 2\left[1 + \cos 2\pi\left(\left(\frac{1}{\lambda_1} - \frac{1}{\lambda_2}\right)z - (v_1 - v_2)t\right)\right] \tag{3.35}$$

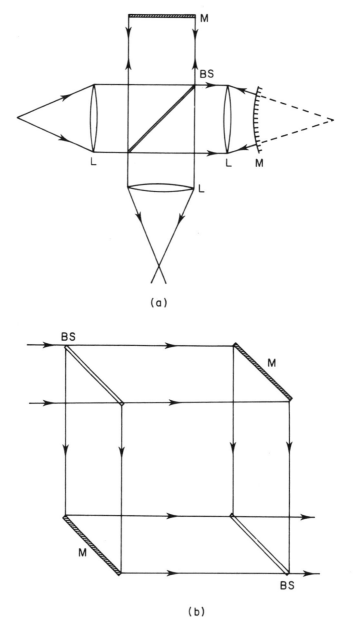

Fig. 3.15. *Examples of amplitude-dividing interferometers. (a) Twyman–Green interferometer. (b) Mach–Zehnder interferometer*

If the frequency difference $v_1 - v_2$ is very small and constant, this variation in I with time can be detected. This is utilized in the dual-frequency Michelson interferometer for length measurement.

Such an interferometer designed by Hewlett-Packard is given in Figure 3.16. The light source is a single-mode laser where the frequency is Zeeman-split into two components f_1

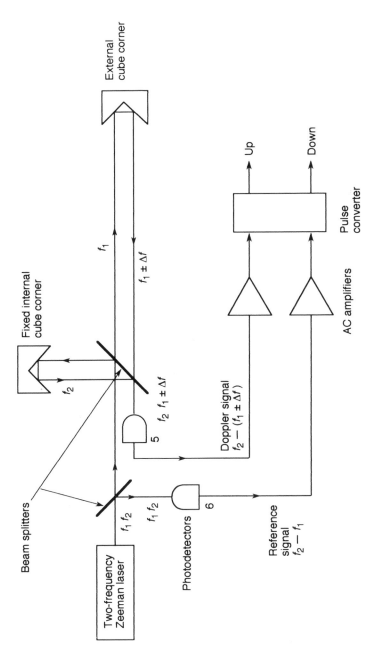

Fig. 3.16.   *The Hewlett-Packard interferometer*

and f_2. (In the figure, frequencies are denoted by f instead of v) with a frequency difference of about 2 MHz. This is achieved by means of an applied magnetic field which splits the light into two orthogonally polarized waves. A polarization-sensitive beamsplitter transmits the component of frequency f_1 to the movable mirror and the component of frequency f_2 to the fixed mirror. These mirrors are cube-corner reflectors in which the reflected beam is parallel to the incoming beam independent of the angle of incidence. If the movable mirror moves with a velocity v_1, the frequency of the reflected light will be Doppler-shifted by an amount $\Delta f = 2v/\lambda_1$. The two light waves interfere on photodetector 5, which therefore gives an electrical signal of frequency $f_2 - (f_1 \pm \Delta f)$ according to Equation (3.35). For two orthogonally polarized waves to interfere, a polarizer has to be placed in front of the detector. A fraction of the laser light of frequencies f_1 and f_2 is sent to the detector 6, producing a reference signal of frequency $f_2 - f_1$. After being amplified, the two electric signals are fed to two frequency counters. The recorded frequencies are then subtracted, giving Δf, which determines the displacement

$$s = vt = \Delta f \lambda t / 2 \tag{3.36}$$

This type of interferometer, also called an a.c. (alternating current) interferometer, gives the displacement in terms of variations in frequency instead of variations in intensity as in the standard Michelson interferometer. Therefore this interferometer is more immune to disturbances like turbulent air in the optical path.

In normal operation the Hewlett-Packard interferometer can measure distances up to 60 m with a resolution of $\lambda/4$ ($\sim 0.16\,\mu m$) which can be extended electronically by factors as high as 100. The accuracy is 5×10^{-7} and the velocity can be measured up to 0.3 m/s. By means of a double cube-corner and an additional beamsplitter, angles and surface flatness can be measured. A new aquisition technique is proposed by Gaal et al. 1993.

3.5.4 Heterodyne (homodyne) detection

The dual-frequency Michelson interferometer is an example of so-called heterodyne detection. This technique is a well-known principle in telecommunication and optical communication. Here the optical signal is mixed with a so-called local oscillator signal at the detector. If we neglect the spatial term (which is constant at the detector), the electric fields of the optical signal ψ_S and of the local oscillator ψ_{LO} are given as

$$\psi_S = U_S \exp[i2\pi(v_S + \phi(t))] \tag{3.37a}$$

$$\psi_{LO} = U_{LO} \exp[i2\pi((v_s + v_{IF})t)] \tag{3.37b}$$

where v_S is the carrier frequency and $\phi(t)$ contains the frequency-modulating information-carrying signal. The local oscillator frequency $v_{LO} = v_S + v_{IF}$ is offset from the carrier frequency by the intermediate frequency v_{IF}. In optical communication v_{IF} is normally in the radio-frequency range from a few tens to hundreds of megahertz. A homodyne detection system results if there is no offset, i.e. $v_{IF} = 0$.

The intensity at the detector becomes

$$I_d = |\psi_S + \psi_{LO}|^2 = U_S^2 + U_{LO}^2 + 2U_S U_{LO} \cos 2\pi(v_{IF}t - \phi(t)) \tag{3.38}$$

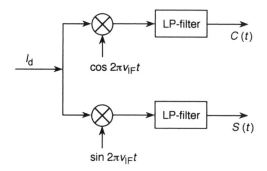

Fig. 3.17.   *The principle of synchronous demodulation*

The signal I_d from the detector can be demodulated by so-called synchronous demodulation consisting of first splitting it into two and multiplying by $\cos 2\pi v_{IF} t$ and $\sin 2\pi v_{IF} t$, see Figure 3.17:

$$I_d \cos 2\pi v_{IF} t = (U_S^2 + U_{LO}^2) \cos 2\pi v_{IF} t + U_S U_{LO} \cos 2\pi (2v_{IF} t - \phi(t)) + U_S U_{LO} \cos 2\pi \phi(t)$$

$$I_d \sin 2\pi v_{IF} t = (U_S^2 + U_{LO}^2) \sin 2\pi v_{IF} t + U_S U_{LO} \sin 2\pi (2v_{IF} t - \phi(t)) + U_S U_{LO} \sin 2\pi \phi(t)$$

$$(3.39)$$

We see that these two expressions both contain two high-frequency terms v_{IF} and $2v_{IF}$. By low-pass filtering of the signals, one is left with

$$C(t) = U_S U_{LO} \cos 2\pi \phi(t) \tag{3.40a}$$

$$S(t) = U_S U_{LO} \sin 2\pi \phi(t) \tag{3.40b}$$

from which the modulating function is given by

$$\phi(t) = \frac{1}{2\pi} \tan^{-1} \left[\frac{S(t)}{C(t)} \right] \tag{3.41}$$

This principle is used in so-called superheterodyne radio receivers. In the dual-frequency Michelson interferometer, $v_{IF} = v_1 - v_2$ and $\phi(t) = 2vt/\lambda_1$. We shall see later that the heterodyne principle is used both in interferometry and moiré, and that the signal frequencies can be in the spatial domain rather than the time domain.

3.6   SPATIAL AND TEMPORAL COHERENCE

After being introduced to interferometers and interferometry, we can consider the topic of coherence a bit further. The following material also demands some knowledge about Fourier transforms, which is treated in Chapter 4 and Appendix B.

In the treatment of the Young's interferometer in Section 3.5.1 we assumed the light to be incident from a point source. Consider the Young set-up in Figure 3.18 where the light is incident from an extended, incoherent, quasimonochromatic light source. In the figure, two light rays from a point P_0 on the source to the point of observation P via the holes P_1 and P_2 in the screen S_1 is drawn. The path-length difference between these two rays is

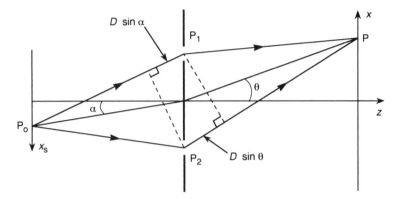

Fig. 3.18. *Young's interferometer with extended light source*

seen to be

$$s = s_1 - s_2 = D \sin \theta - D \sin \alpha = D(\theta - \alpha) \tag{3.42}$$

where in the last equality we have approximated the sines by the angles.

Assume that the intensity at P when only P_1 is open is $I(\alpha)/2$. Then, because of the small path-length difference, the intensity at P when only P_2 is open also will be $I(\alpha)/2$. The intensity at P due to the light from P_0 therefore can be written as

$$\Delta I = I(\alpha)\{1 + \cos[kD(\theta - \alpha)]\} \tag{3.43}$$

To find the total intensity at P we have to integrate Equation (3.43) over the range of α from $-\alpha_m/2$ to $\alpha_m/2$. Since $I(\alpha)$ is zero outside the source, we can change the integration limits to $-\infty, \infty$. We then get

$$I = \int_{-\infty}^{\infty} \Delta I \, d\alpha = \int_{-\infty}^{\infty} I(\alpha) \, d\alpha + \int_{-\infty}^{\infty} I(\alpha) \cos[kD(\theta - \alpha)] \, d\alpha$$

$$= I_t \left\{ 1 + \frac{1}{I_t} \int_{-\infty}^{\infty} I(\alpha) \cos[kD(\theta - \alpha)] \, d\alpha \right\} \tag{3.44}$$

where

$$I_t = \int_{-\infty}^{\infty} I(\alpha) \, d\alpha \tag{3.45}$$

is the total intensity with no interference.

The integrand in the last integral can be rewritten to read

$$I(\alpha) \cos[kD(\theta - \alpha)] = I(\alpha) \operatorname{Re}\{e^{ikD(\theta - \alpha)}\} = \operatorname{Re}\{e^{ikD\theta} I(\alpha) e^{-ikD\alpha}\} \tag{3.46}$$

This, put into the last integral of Equation (3.44), gives

$$\frac{1}{I_t} \int_{-\infty}^{\infty} \operatorname{Re}\{e^{ikD\theta} I(\alpha) e^{-ikD\alpha}\} \, d\alpha = \frac{1}{I_t} \operatorname{Re}\left\{ e^{ikD\theta} \int_{-\infty}^{\infty} I(\alpha) e^{-ikD\alpha} d\alpha \right\}$$

$$= \operatorname{Re}\{e^{ikD\theta} \gamma_{12}\} = |\gamma_{12}| \cos(kD\theta + \psi) \tag{3.47}$$

where

$$\gamma_{12} = |\gamma_{12}|e^{i\psi} = \frac{\displaystyle\int_{-\infty}^{\infty} I(\alpha)\, e^{ikD\alpha}\, d\alpha}{\displaystyle\int_{-\infty}^{\infty} I(\alpha)\, d\alpha} \qquad (3.48)$$

This, inserted into Equation (3.44), gives

$$I(P) = I_t\{1 + |\gamma_{12}|\cos[kD\theta + \psi]\} \qquad (3.49)$$

From the discussion in Section 3.2 we recognize γ_{12} as the complex degree of spatial coherence. We see that γ_{12} is equal to the normalized Fourier transform of the intensity distribution of the source, with the frequency coordinate equal to D/λ.

For a square disc of uniform intensity

$$I(\alpha) = \frac{I_t}{\Delta\alpha}\, \mathrm{rect}\left(\frac{\alpha}{\Delta\alpha}\right) \qquad (3.50)$$

we get

$$|\gamma_{12}| = \left|\mathrm{sinc}\left(\frac{D\Delta\alpha}{\lambda}\right)\right| \qquad (3.51)$$

from which we find that the visibility has its first zero for

$$D = D_c = \frac{\lambda}{\Delta\alpha} \qquad (3.52)$$

For a circular disc of uniform intensity

$$I(\alpha) = \frac{I_t}{\Delta\alpha}\, \mathrm{circ}\left(\frac{\alpha}{\Delta\alpha}\right) \qquad (3.53)$$

we get

$$|\gamma_{12}| = \left|\frac{J_1\left(\dfrac{\pi D\Delta\alpha}{\lambda}\right)}{\dfrac{\pi D\Delta\alpha}{\lambda}}\right| \qquad (3.54)$$

with the first zero occurring for

$$D = D_c = 1.22\frac{\lambda}{\Delta\alpha} \qquad (3.55)$$

In an experiment conducted by A. A. Michelson at the Mt. Wilson Laboratory, California in December 1920, the fringes formed in the stellar interferometer (see Figure 3.13c) by the light from the star Betelgeuse were made to vanish at $D = 121$ inches (3070 mm). With $\lambda = 570$ nm, $\Delta\alpha = 22.6 \times 10^{-8}$ rad, and from its known distance determined from parallax measurements, the star's diameter turned out to be about 380 million km or roughly 280 times that of the Sun.

In the description of the Michelson interferometer in Section 3.5.2 we assumed the light source to be monochromatic and found the output intensity to be (assuming a 50/50 beamsplitter):

$$I = \Delta I\left(1 + \cos\left(\frac{2\pi v d}{c}\right)\right) = \Delta I(1 + \cos(2\pi v \tau)) \tag{3.56}$$

where ΔI is the total intensity of the incoming beam, d is the path-length difference and $\tau = d/c$ is the time difference between the two paths. When using a light source of N distinct frequencies, each of intensity I_n, we will, since each component can interfere with itself only, get an output intensity equal to

$$I = \sum_n I_n(1 + \cos(2\pi v_n \tau)) \tag{3.57}$$

Ordinary light sources have a continuous distribution of frequencies $I(v)$, in which case the sum in Equation (3.57) is converted into the integral

$$I = \int_0^\infty I(v)(1 + \cos(2\pi v \tau))\,dv = I_0\left[1 + \int_0^\infty P(v)\cos(2\pi v \tau)dv\right] \tag{3.58}$$

where

$$I_0 = \int_0^\infty I(v)\,dv \tag{3.59}$$

is the total intensity and

$$P(v) = \frac{I(v)}{I_0} \tag{3.60}$$

is known as the normalized spectral distribution function of the source. Since light frequencies are positive, the lower integration limit of Equation (3.58) is zero. However, if the spectral distribution is peaked about $v = v_0$, we can write

$$P(v) = S(v - v_0) \tag{3.61}$$

which gives

$$\int_0^\infty P(v)\cos(2\pi v \tau)dv = \mathrm{Re}\left\{\int_{-\infty}^\infty S(v - v_0)e^{-i2\pi v \tau}dv\right\}$$

$$= \mathrm{Re}\left\{e^{-i2\pi v_0 \tau}\int_{-\infty}^\infty S(v)e^{-i2\pi v \tau}dv\right\} = |\gamma(\tau)|\cos(2\pi v_0 \tau) \tag{3.62}$$

This, put into Equation (3.58), gives

$$I = I_0[1 + |\gamma(\tau)|\cos(2\pi v_0 \tau)] \tag{3.63}$$

where

$$\gamma(\tau) = \int_{-\infty}^\infty S(v)e^{-i2\pi v \tau}dv \tag{3.64}$$

is recognized as the complex temporal degree of coherence and is equal to the Fourier transform of the spectral distribution function of the source with the origin of the frequency axis moved to v_0.

In the development of the spatial degree of coherence, we assumed a quasimonochromatic source and in the development of the temporal degree of coherence we assumed a plane wave. Coherence theory can however be developed without these assumptions, and it can be shown that we can introduce a generalized coherence function

$$\gamma_{12}(\tau, D) \approx \gamma_{12}(\tau, 0)\gamma_{12}(0, D) = \gamma(\tau)\gamma_{12} \qquad (3.65)$$

This equation is known as the reduction property of the complex degree of coherence. In the Young interferometer experiment, we can in principle measure $\gamma(\tau, D)$, but it is difficult, due to the limited region of overlap between the waves from P_1 and P_2 to obtain τ much different from zero.

PROBLEMS

3.1 Find the resultant $U = U_1 + U_2$, where $U_1 = U_0 \exp i(kz + \omega t)$ and $U_2 = U_0 \exp i(kz - \omega t)$. Describe the resultant wave.

3.2 A carrier wave of frequency ω_c is amplitude modulated by a sine wave of frequency ω_m, that is

$$U = U_0(1 + \alpha \cos \omega_m t) e^{i\omega_c t}$$

Show that this is equivalent to the superposition of three waves of frequencies $\omega_c, \omega_c + \omega_m$, and $\omega_c - \omega_m$. When a number of modulating frequencies are present, we write U as a Fourier series and sum over all values of ω_m. The terms $\omega_c + \omega_m$ constitute what is called the upper sideband while all of the $\omega_c - \omega_m$ terms form the lower sideband. What bandwidth would you need in order to transmit the complete audible range?

3.3 Calculate the coherence length of the following sources:

(a) filtered thermic radiation of bandwidth 1 nm at wavelength 600 nm;
(b) a multimode He–Ne laser with bandwidth 1 GHz;
(c) a monomode He–Ne laser with bandwidth 10 kHz.

3.4(a) In a magnetic-field technique an He–Ne laser can be stabilized to 2 parts in 10^{10}. At 632.8 nm, what would be the coherence length of a laser with such a frequency stability?
(b) Imagine that we chop a continuous laser beam (assumed to be monochromatic at $\lambda_0 = 632.8$ nm) into 0.1 ns pulses using some sort of shutter. Compute the resultant linewidth $\Delta\lambda$, bandwidth and coherence length. Find the bandwidth and linewidth which would result if we could chop at 10^{15} Hz.

3.5 In a Lloyd's mirror experiment with light of wavelength $= 500$ nm, it is found that the bright fringes on a screen 1 m away from the source are 1 mm apart. Calculate the (perpendicular) height of the source slit from the mirror.

3.6 Show that a for the Fresnel biprism of Figure P3.1 is given by $a = 2d(n-1)\alpha$.

3.7 The Fresnel biprism is used to obtain fringes from a point source which is placed 2 m from the screen and the prism is midway between the source and the screen. Let the light have a wavelength $\lambda_0 = 500$ nm, and the index of refraction of the glass be $n = 1.5$. What is the prism angle if the separation of the fringes is 0.5 mm?

3.8 As mentioned in the text, the spatial coherence function γ_{12} is equal to the Fourier transform of the intensity distribution of the source.

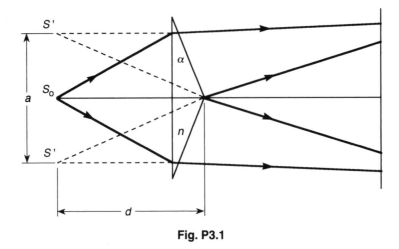

Fig. P3.1

Take the Sun as the light source. For simplicity, assume that the Sun is square and that its intensity distribution is constant. The angle subtended by the Sun is $\Delta\alpha = 9.3$ millirad ($\approx 0.53°$).

(a) Find the spatial degree of coherence $|\gamma_{12}|$ expressed by the spatial separation D, $\Delta\alpha$ and λ.

(b) Find the spatial coherence length D_c and the coherence area $\pi(D_c/2)^2$. Use $\lambda_0 = 500$ nm.

Next assume that we use the reflected sunlight from a window of area $1\,\text{m} \times 1\,\text{m}$ a distance 1 km away as the light source.

(c) Find the spatial coherence length and area in this case.

3.9 Consider a light source consisting of two point sources with an angular separation equal to $\Delta\alpha$. Repeat (a) and (b) of Problem 3.8 and compare.

4 Diffraction

4.1 INTRODUCTION

Figure 4.1 shows a plane wave which is partly blocked by a screen S_1 before falling onto a screen S_2. According to geometrical optics, a sharp edge is formed by the shadow at point A. By closer inspection, however, one finds that this is not strictly correct. The light distribution is not sharply bounded, but forms a pattern in a small region around A. This must be due to a bending of the light around the edge of S_1. This bending is called diffraction and the light pattern seen on S_2 as a result of interference between the bent light waves is called a diffraction pattern.

Another example of this phenomenon can be observed by sending light through a small hole. If this hole is made small enough, the light will not propagate as a narrow beam but as a spherical wave from the centre of the hole (see Figure 4.2). This is evidence of Huygens' principle which says that every point on a wavefront can be regarded as a source of secondary spherical wavelets. By adding these wavelets and calculating the intensity distribution over a given plane, one finds the diffraction pattern in that plane. This simple

Fig. 4.1

Fig. 4.2

principle has proved to be very fruitful and constitutes the foundation of the classical diffraction theory.

4.2 DIFFRACTION FROM A SINGLE SLIT

With this simple assumption, we shall try to calculate the diffraction pattern from a long, narrow slit (see Figure 4.3). The slit width a in the x_0-direction is much smaller than the length in the y_0-direction. We therefore consider the problem as one-dimensional. From the left, a plane wave with unit amplitude is falling normally onto the slit. According to Huygens' principle, the contribution to the field at a point x from an arbitrary point x_0 inside the slit is equal to the field of a spherical wave with its centre at x_0. We assume that the distance z to the observation plane is so long that the field at x can be regarded as a plane wave. The expression for a plane wave, obliquely incident at an angle θ, was derived in Equation (1.9):

$$\Delta u(x) = \exp(ikx \sin \theta - i\phi) = e^{i\alpha x} e^{-i\phi} \tag{4.1}$$

where we have put

$$\alpha = k \sin \theta \simeq k \frac{x}{z} = \frac{2\pi x}{\lambda z} \tag{4.2}$$

We put the arbitrary phase constant ϕ equal to zero for the wavelet originating from $x_0 = 0$. The relative phase from any point x_0 inside the slit then becomes (see Figure 4.3)

$$\phi = kx_0 \sin \theta = \alpha x_0$$

To calculate the total field at the point x, we have to sum the Huygens' wavelets from all points inside the slit. Therefore we must integrate Equation (4.1) over the slit width:

$$u(x) = e^{i\alpha x} \int_{-a/2}^{a/2} e^{-i\alpha x_0} dx_0 = e^{i\alpha x} \frac{a \sin(\alpha a/2)}{\alpha a/2} \tag{4.3}$$

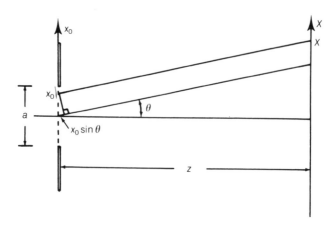

Fig. 4.3

The intensity becomes

$$I(x) = |u(x)|^2 = a^2 \frac{\sin^2\left(\pi\frac{ax}{\lambda z}\right)}{\left(\pi\frac{ax}{\lambda z}\right)^2} \qquad (4.4)$$

where we have inserted the expression for α from Equation (4.2).

In deriving Equation (4.4), we have made some approximations. These are called the Fraunhofer approximation in optics. To fulfil this approximation, the observation plane must in general be moved far away from the diffracting object. A simple way of fulfilling this condition is to observe the diffraction pattern in the focal plane of a lens (see Section 4.3.2).

In Figure 4.4, the Fraunhofer diffraction pattern from a single slit according to Equation (4.4) is given. The distribution constitutes a pattern of light and dark fringes. From Equation (4.4) we find that the distance between n minima is equal to

$$\Delta x = n\frac{\lambda z}{a} \qquad (4.5)$$

We see that Δx is inversely proportional to the slit width a. This can be utilized for measuring a by measuring the distance Δx, for example, by means of photodetectors.

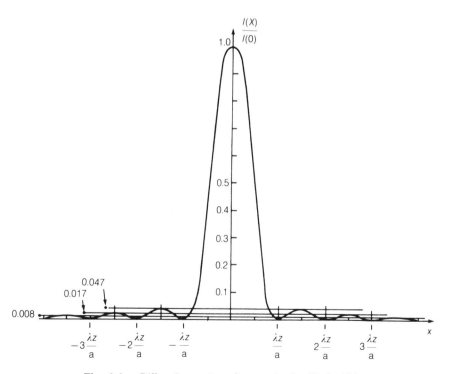

Fig. 4.4. *Diffraction pattern from a single slit of width a*

Note that the diffraction pattern from an opaque strip of width a will be the same as from a slit of the same width. The above arguments can therefore be applied to, for example, the measurement of wire thicknesses. The signals from the photodetectors can be fed to the wire-drawing machine for automatic regulation of the wire-drawing process.

4.3 DIFFRACTION FROM A GRATING

4.3.1 The grating equation. Amplitude transmittance

Figure 4.5 shows a plane wave normally incident on a grating with a grating period equal to d. The grating lines are so narrow that we can regard the light from each opening as cylindrical waves. In Figure 4.5(b) we have drawn three of these openings, A, B and C, each with five concentric circles separated by λ representing the cylindrical waves. The tangent to circle number 5 for all openings will represent a plane wave propagating in the z-direction.

The tangent to circle 5 from opening A, circle 4 from B and circle number 3 from C will represent a plane wave propagating in a direction making an angle θ_1 to the z-axis. From the figure we see that

$$\sin \theta_1 = \lambda/d$$

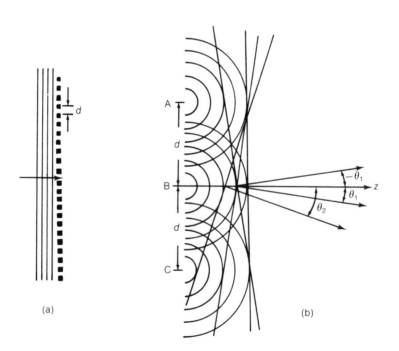

Fig. 4.5. Diffraction from a square wave grating

Fig. 4.6. *Amplitude transmittance t of a square wave grating*

The tangent to circle 5 from opening A, circle 3 from B and circle 1 from C will represent a plane wave propagating in a direction making an angle θ_2 to the z-axis given by

$$\sin \theta_2 = 2\lambda/d$$

In the same manner we can proceed up to the plane wave number n making an angle θ_n to the z-axis given by

$$\sin \theta_n = n\lambda/d \qquad (4.6)$$

Equation (4.6) is called the grating equation. Also in the same manner we can draw the tangent to circle 5 from opening C, circle 4 from B and circle 3 from A and so on. Therefore n in Equation (4.6) will be an integer between $-\infty$ and $+\infty$.

The grating in Figure 4.5 can be represented by the function $t(x)$ in Figure 4.6. This is a square-wave function discontinuously varying between 0 and 1. If the wave incident on the grating is represented by u_i, the wave just behind the grating is given by

$$u_u = t(x)u_i \qquad (4.7)$$

Therefore, behind the grating plane $u_u = u_i$ where $t(x) = 1$, i.e. the light is transmitted and $u_u = 0$ wherever $t(x) = 0$, i.e. the light is blocked.

The function $t(x)$ is called the complex amplitude transmittance of the grating. We have seen that such a grating will diffract plane waves in directions given by Equation (4.6). If we turned the propagation direction 180° around for all these waves, it should not be difficult to imagine that they would interfere, forming an interference pattern with a light distribution given by $t(x)$ in Figure 4.6. In the same way we realize that a sinusoidal grating (which can be formed on a photographic film by interference between two plane waves) will diffract two plane waves propagating symmetrically around the z-axis when illuminated by a plane wave like the square wave grating in Figure 4.5. Diffraction from a sinusoidal (consinusoidal) grating is therefore also described by Equation (4.6), but now n will assume the values $-1, 0$ and 1 only.

Further reasoning along the same lines tells us that a zone-plate pattern formed by registration (e.g. on a photographic film) of the interference between a plane wave and a spherical wave (see Figure 4.7(a)), will diffract two spherical waves. One of them will be a diverging spherical wave with its centre at P and the other will converge (focus) to a point P′ separated from the zone-plate by the same distance a as the point P (see Figure 4.7(b)). These arguments are perhaps not so easy to accept, but are nevertheless correct.

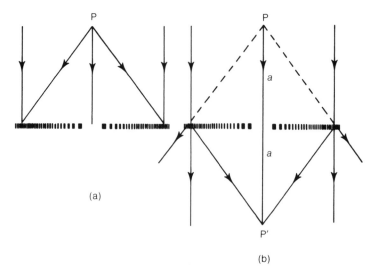

(a)

(b)

Fig. 4.7. *(a) Zone plate formed by interference between a spherical and a plane wave. (b) Diffraction of a plane wave from a zone plate*

4.3.2 The spatial frequency spectrum

Assume that we place a positive lens behind the grating in Figure 4.5 such as in Figure 4.8. In Section 1.9 (Equation (1.20)) we have shown that a plane wave in the xz-plane with propagation direction an angle θ to the optical axis (the z-axis) will focus to a point in the focal plane of the lens at a distance x_f from the z-axis given by

$$x_f = f \tan \theta \tag{4.8}$$

where f is the focal length.

By substituting Equation (4.6) we get

$$x_f = n \frac{\lambda f}{d} = n \lambda f f_0 \tag{4.9}$$

Fig. 4.8

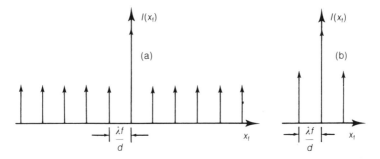

Fig. 4.9. *Diffraction patterns from (a) square wave grating, (b) sinusoidal grating*

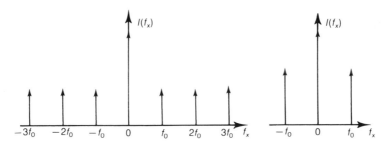

Fig. 4.10. *Spatial frequency spectra from (a) square wave grating, (b) sinusoidal grating*

where we have used the approximation $\sin \theta = \tan \theta$ and inserted the grating frequency $f_0 = 1/d$. If we represent the intensity distribution in a focal point by an arrow, the intensity distribution in a focal plane in Figure 4.8 will be like that given in Figure 4.9(a). By exchanging the square-wave grating with a sinusoidal grating, the intensity distribution in the focal plane will be like that given in Figure 4.9(b).

Figure 4.10 is a reproduction of Figure 4.9 apart from a rescaling of the ordinate axis from x_f of dimension length to $f_x = x_f/\lambda f$ of dimension inverse length, i.e. spatial frequency. In that way we get a direct representation of the frequency content or plane-wave content of the gratings. We see that the sinusoidal grating contains the frequencies $\pm f_0$ and 0, while the square-wave grating contains all positive and negative integer multiples of f_0. The diagrams in Figure 4.10 are called spatial frequency spectra.

If we successively put into the set-up in Figure 4.8 sinusoidal gratings of frequencies $f_0, 2f_0, 3f_0, \dots, nf_0$, and if we could add all the resulting spectra, we would get a spectrum like that given in Figure 4.10(a). This would be a proof of the fact that a square-wave grating can be represented by a sum of sinusoidal (cosinusoidal) gratings of frequencies which are integer multiples of the basic frequency f_0, in other words a Fourier series. This is further evidenced in Figure 4.11 where, in Figure 4.11(c) we see that the approximation to a square-wave grating is already quite good by adding the four first terms of the series. To improve the reproduction of the edges of the square wave grating, one has to include the higher-order terms of the series. Sharp edges in an object will therefore represent high spatial frequencies.

4.4 FOURIER OPTICS

Let us turn back to Section 4.2 where we found an expression for the field $u(x_f)$ in the x_f-plane diffracted from a single slit of width a in the x-plane at a distance z (see Equation (4.3)):

$$u(x_f) = \int_{-a/2}^{a/2} e^{-i2\pi f_x x} dx \tag{4.10}$$

where $f_x = x_f/\lambda z$ and where we have omitted a constant phase factor. The transmittance function for a single slit would be

$$t(x) = \begin{cases} 1 & \text{for } |x| < a/2 \\ 0 & \text{otherwise} \end{cases} \tag{4.11}$$

By putting $t(x)$ into the integral of Equation (4.10) we may let the limits of integration approach $\pm\infty$, and we get

$$u(x_f) = \int_{-\infty}^{\infty} t(x) e^{-i2\pi f_x x} dx \tag{4.12}$$

This is a Fourier integral and $u(x_f)$ is called the Fourier transform of $t(x)$. In the general case where the transmittance function varies both in the x- and y-direction, we get

$$u(x_f, y_f) = \int_{-\infty}^{\infty} \int t(x, y) e^{-i2\pi(f_x x + f_y y)} dx \, dy \tag{4.13}$$

which in shorthand notation can be written

$$u(x_f, y_f) = T(f_x, f_y) = \mathscr{F}\{t(x, y)\} \tag{4.14}$$

where $\mathscr{F}\{t(x, y)\}$ means 'the Fourier transform of $t(x, y)$'.

In deriving Equation (4.3) we assumed a plane wave of unit amplitude incident on the slit. If a light wave given by $u_i(x, y)$ falls onto an object given by the transmittance function $t(x, y)$, the field just behind the object is $u(x, y) = t(x, y)u_i(x, y)$ and the field in the x_f-plane becomes

$$u(x_f, y_f) = \frac{K}{i\lambda z} \mathscr{F}\{u(x, y)\} \tag{4.15}$$

Here K is a pure phase factor ($|K|^2 = 1$) which is unimportant when calculating the intensity. By the factor $(1/i\lambda z)$ we have brought Equation (4.15) into accordance with the Huygens–Fresnel diffraction theory.

As mentioned in Section 4.2, the approximations leading to Equation (4.3) and therefore Equation (4.15) are called the Fraunhofer approximation. To fulfil this, the plane of observation has to be far away from the object. A more practical way of fulfilling this requirement is to place the plane of observation in the focal plane of a lens as in Figure 4.8. z in Equation (4.15) then has to be replaced by the focal length f. Other practical methods are treated in Section 4.5.1. We also mention that by placing the object in the front focal plane (to the left of the lens) in Figure 4.8, we have $K = 1$, and we get a direct Fourier transform.

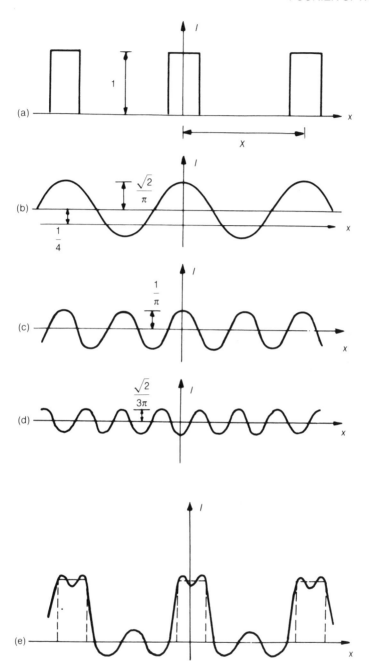

Fig. 4.11.   *Fourier decomposition of a square wave grating. (a) The transmittance function of the grating. (b) The constant term and the first harmonic of the Fourier series. (c) The second harmonic. (d) The third harmonic. (e) The sum of the four first terms of the series. The transmittance function of the grating is shown dashed*

The way we have derived the general formula of Equation (4.15) is of course by no means a strict proof of its validity. Rigorous diffraction theory using the same approximations leads, however, to the same result. Equation (4.15) is a powerful tool in calculating diffraction patterns and analysis of optical systems. Some of its consequences are treated more extensively in Appendix B.

For example, the calculation of the frequency spectrum of a sinusoidal grating given by

$$t(x) = 1 + \cos 2\pi f_0 x \tag{4.16}$$

now becomes straightforward. By using Equation (4.13) we get

$$
\begin{aligned}
u(x_f) &= \int_{-\infty}^{\infty} (1 + \cos 2\pi f_0 x) e^{-i2\pi f_x x} dx \\
&= \int_{-\infty}^{\infty} (1 + \tfrac{1}{2} e^{i2\pi f_0 x} + \tfrac{1}{2} e^{-i2\pi f_0 x}) e^{-i2\pi f_x x} dx \\
&= \int_{-\infty}^{\infty} (e^{-i2\pi f_x x} + \tfrac{1}{2} e^{-i2\pi (f_x - f_0) x} + \tfrac{1}{2} e^{-i2\pi (f_x - f_0) x}) dx \\
&= \delta(f_x) + \tfrac{1}{2} \delta(f_x - f_0) + \tfrac{1}{2} \delta(f_x + f_0)
\end{aligned}
\tag{4.17}
$$

The last equality follows from the definition of the delta function given in Equation (B.11) in Appendix B.2. Equation (4.17) shows that the spectrum of a sinusoidal grating is given by the three delta functions, i.e. three focal points. These are the zero order at $f_x = 0$ and the two side orders at $f_x = \pm f_0$ (see Figure 4.9(b)).

4.5 OPTICAL FILTERING

Figure 4.12 shows a point source (1) placed in the focal plane of a lens (2) resulting in a plane wave falling onto a square wave grating (3) which lies in the object plane. A lens (4) placed a distance a from the object plane, images the square wave grating on to the image plane where the intensity distribution (5) of the image of the grating can be observed.

Although the figure shows that only the $\pm$1st side orders are accepted by the lens (4), we will in the following assume that all plane-wave components diffracted from the grating will go through the lens. We can therefore, in the same way as in Figure 4.8, observe the spectrum of the grating in the back focal plane of the lens.

Let us now consider two cases:

Case 1. We place a square-wave grating of basic frequency f_0 in the object plane. The distance between the focal points in the focal plane then becomes $\lambda f f_0$ (see Equation (4.9) and Figure 4.9). In the image plane we will see a square-wave grating of basic frequency f_i where $1/f_i = m(1/f_0)$ and $m = b/a$ is the magnification from the object plane to the image plane.

Case 2. We place a square wave grating of basic frequency $2f_0$ in the object plane. The distance between the focal points now becomes $2\lambda f f_0$ and the basic frequency of the imaged grating will be $2f_i$.

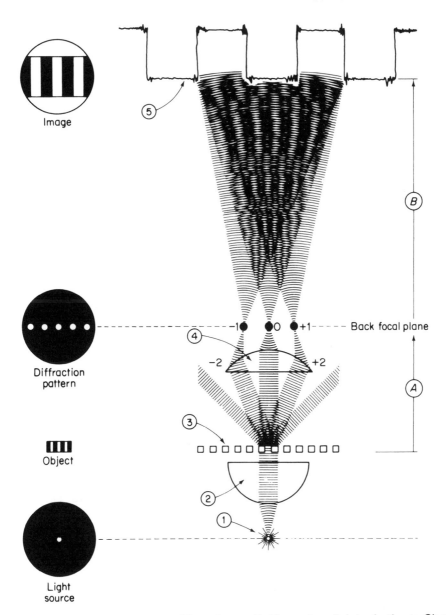

Image

Diffraction pattern

Object

Light source

Back focal plane

Fig. 4.12. *Optical filtering process. (From Jurgen R. Meyer-Arendt, Introduction to Classical and Modern Optics, © 1972, p. 393. Reprinted by permission of Prentice-Hall, Inc., Englewood Cliffs, New Jersey)*

If in case 1 we placed in the focal plane a screen with holes separated by the distance $2\lambda f f_0$ and adjusted it until every second focal point in the spectrum was let through, then the situation in case 2 would be simulated. In other words, a grating of basic frequency f_0 in the object plane would have resulted in a grating of basic frequency $2f_i$ in the image plane. Such a manipulation of the spectrum is called optical filtering and the back focal plane of lens (4) is called the filter plane. What we have here described is only one of many examples of optical filtering. The types of filters can have many variations and can be rather more complicated than a screen with holes. We shall in Section 4.7 consider a special type of filtering which we later will apply in practical problems.

4.5.1 Practical filtering set-ups

In Section 4.3.1 we derived the grating equation

$$d \sin \theta_n = n\lambda \tag{4.18}$$

which applies to normal incidence on the grating. This equation implies that the phase difference between light from the different openings in the grating must be an integer number of wavelengths. When the light is obliquely incident upon the grating at an angle θ_i, this condition is fulfilled by the general grating equation:

$$d(\sin \theta_i + \sin \theta_n) = n\lambda \tag{4.19}$$

This is illustrated in Figure 4.13.

Figure 4.14 shows the same set-up as in Figure 4.8 apart from the grating being moved to the other side of the lens a distance s from the back focal plane. Assume that the first side order from an arbitrary point x on the grating is diffracted to the point x_f. The grating equation (4.19) applied to this light path gives

$$d(\sin \theta_i + \sin \theta_1) \simeq d(\tan \theta_i + \tan \theta_1) = d\left(\frac{x}{s} + \frac{x_f - x}{s}\right) = \lambda \tag{4.20}$$

Equation (4.20) implies that the first side order will be diffracted to the same point

$$x_f = \frac{\lambda s}{d} \tag{4.21}$$

in the focal plane independent of x, i.e. from all points on the grating. The intensity distribution in the focal plane therefore becomes like that in Figure 4.9 except that the diffraction orders now are separated by a distance $\lambda s/d$.

Fig. 4.13

Fig. 4.14

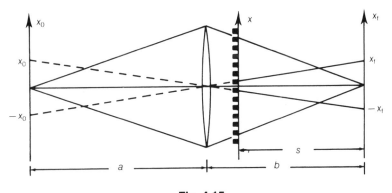

Fig. 4.15

The above arguments do not assume a plane wave incident on the lens. The result remains the same when a point source is imaged by the lens on to the optical axis as in Figure 4.15.

Our next question is: how will the light be distributed in the x_0-plane with a point source at $x_f = 0$ in the set-up of Figure 4.15? It should be easy to accept that the answer is given by removing the grating and placing two point sources at the positions $\pm x_f = \pm \lambda s/d$.

The first two diffraction orders are therefore found at the points $\pm x_0 = \pm (a/b)\lambda s/d$.

The conclusions of the above considerations are collected in Figure 4.16. Here, P_i is the image point of the point source P_0 on the optical axis. They are separated by distances given by the lens formula

$$\frac{1}{a} + \frac{1}{b} = \frac{1}{f} \tag{4.22}$$

where $f = $ the lens focal length.

The important point to note is that independent of the positioning of the grating (or another transparent object) in the light path between P_0 and P_i, the spectrum will be found in the x_f-plane. The distance between the diffraction orders becomes: if we place the grating

Fig. 4.16

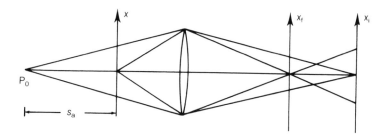

Fig. 4.17. *Optical filtering set-up with one single lens*

(1) to the right of the lens at a distance s_b from P_i

$$x_f = \lambda s_b / d \tag{4.23}$$

(2) to the left of the lens at a distance s_a from P_0

$$x_f = (b/a)\lambda s_a/d = \frac{a - t_a}{a - f} \frac{\lambda f}{d} \tag{4.24}$$

where t_a is the distance from the grating to the lens.

These results make possible a lot of different filtering set-ups. Figure 4.17 shows the simplest of them all. Here the lens works as both the transforming and the imaging lens. The object in the x-plane is imaged to the x_i-plane while the filter plane is in the x_f-plane, the image plane of the point source P_0.

Figure 4.18 shows a practical filtering set-up often used for optical filtering of moiré and speckle photographs. The film is placed just to the right of the lens a distance s_b from the filter plane. By imaging the film through a hole in the filter plane a distance x_f from the optical axis, one is filtering out the first side order of a grating of frequency

$$f_x = \frac{x_f}{\lambda s_b} \tag{4.25}$$

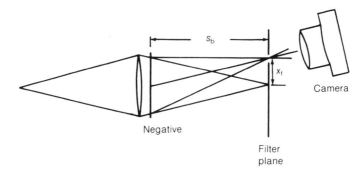

Fig. 4.18. *Practical filtering set-up*

Owing, for example, to a deformation, this grating can be regarded as phase-modulated, and the side order therefore has been broadened, see Section 4.7. When the hole in the filter plane is made wide enough to let through this modulated side order, the image of the object on the film will be covered by moiré or speckle Fourier fringes. This will be treated in more detail in Chapters 7 and 8.

4.6 PHYSICAL OPTICS DESCRIPTION OF IMAGE FORMATION

We are now in a position to look more closely at a lens system from a physical optics point of view. In Section 4.5 we have shown that by placing a transparent diffracting object anywhere in the light path between P_0 and P_i of Figure 4.16, we obtain the spectrum (i.e the Fourier transform) of the object in the x_f-plane. But this argument applies also to the circular lens aperture itself. The Fourier transform of a circular opening given by (see definition, Equation (B.17) in Appendix B.2)

$$\mathrm{circ}\left(\frac{r}{D/2}\right) \tag{4.26}$$

is given in Table B.1 (Appendix B.2) as

$$\mathscr{F}\left\{\mathrm{circ}\left(\frac{r}{D/2}\right)\right\} = \left(\frac{D}{2}\right)^2 \frac{J_1(\pi D\rho)}{D\rho/2} \tag{4.27}$$

where D is the diameter of the lens, J_1 is the first-order Bessel function and where we also have made use of the similarity theorem, Equation (B.2b), Appendix B.1. The frequency coordinate in this case is equal to

$$\rho = \frac{r_f}{\lambda b} \tag{4.28}$$

where b is the image distance and

$$r_f = \sqrt{x_f^2 + y_f^2} \tag{4.29}$$

The intensity distribution in the x_f, y_f-plane in Figure 4.16 with a point source at P_0 therefore becomes proportional to

$$I(r_f) = \left[\frac{J_1(kDr_f/2b)}{(kDr_f/2b)} \right]^2 \tag{4.30}$$

This intensity distribution is generally referred to as the Airy pattern, after G. B. Airy who first derived it. Figure 4.19(a) shows a cross-section and Figure 4.19(b) shows a photograph of this Airy pattern. The first minimum of $J_1(\pi x)$ occurs for $x = 1.22$ which gives for the radius of the so-called Airy disc

$$\Delta r_f = 1.22 \frac{\lambda b}{D} \tag{4.31}$$

This result shows that even for an aberration-free system, an object point is not imaged as an image point predicted by geometrical optics, but as an intensity distribution given by Equation (4.30).

An imaging system is usually composed, not of a single lens, but of several lenses, perhaps some positive and some negative. To specify the properties of a lens system we consider it as a 'black box' with the entrance and exit pupils as 'terminals'. The entrance pupil represents a finite aperture (effective or real) through which light must pass to reach the imaging elements and the exit pupil represents a finite aperture through which the light

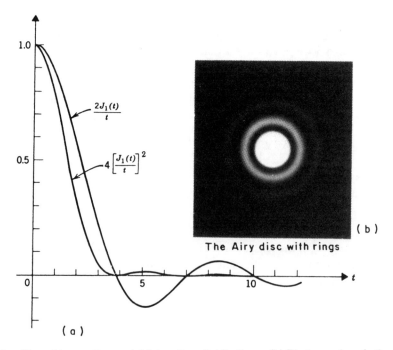

Fig. 4.19. *The Airy pattern. (a) Intensity distribution. (b) Photograph of the pattern. (Reproduced from Klein 1970 by permission of John Wiley & Sons Inc)*

must pass as it leaves the imaging elements on its way to the image plane. It is usually assumed that passage of light between the entrance and exit planes is adequately described by geometrical optics. To find the two pupils, one first has to identify the smallest (physical) aperture (the aperture stop) within the system. Then the entrance pupil is found by imaging (geometrically) this aperture through all the lens elements to the left (towards the object plane) and the exit pupil by imaging the aperture through the lens elements to the right, see Section 2.4. The lens system is said to be diffraction limited if a diverging spherical wave incident on the entrance pupil is mapped into a converging spherical wave at the exit pupil.

With this model of the imaging system, it is possible to associate all diffraction limitations with either the light propagation from object to entrance pupil or from the exit pupil to the image. In fact, the two points of view are entirely equivalent.

The view that regards diffraction effects as resulting from the finite entrance pupil was first examined by Ernst Abbe in 1873. According to his theory, only a limited number of diffracted orders from an object are intercepted by the finite entrance pupil (see e.g. Figure 4.12). The equivalent view of regarding diffraction effects as resulting from the finite exit pupil was presented by Lord Rayleigh in 1896. We will adopt this viewpoint here.

4.6.1 The coherent transfer function

Generally, if the exit pupil is represented not merely by a circ-function but by some general aperture function $P(x, y)$, the field amplitude distribution in the image plane due to an on-axis point source in the object plane will be given by a function $h(x_i, y_i)$

$$h(x_i, y_i) = \mathcal{F}\{P(x, y)\} \tag{4.32}$$

Equation (4.27) is a special case of this formula. Here x, y are lens aperture coordinates, x_i, y_i are image plane coordinates and the Fourier frequency coordinates are given by $f_x = x_i/\lambda b, f_y = y_i/\lambda b$. With the object point moved off-axis to a point x_o, y_o in the object plane, the resulting field distribution in the image plane will be centred around a point of coordinates $x_i = -mx_o, y_i = -my_o$ where m is the transversal magnification. If the field amplitude of this object point is given by $\Delta u_o(x_o, y_o)$ (which ideally should be a delta-function $\delta(x_o - x'_o, y_o - y'_o)$), the field amplitude distribution in the image plane therefore is given by

$$\Delta u_i(x_i, y_i) = h(x_i - mx_o, y_i - my_o)\Delta u_o(x_o, y_o) \tag{4.33}$$

If, in the object plane there are many point sources which are mutually coherent, we have to add the contribution from each point according to Equation (4.33). Moreover, if in the object plane there is a continuous and coherent field amplitude distribution given by $u_o(x_o, y_o)$, this sum will be converted to an integral, viz.

$$u_i(x_i, y_i) = \int\int_{-\infty}^{\infty} h(x_i - mx_o, y_i - mx_o)u_o(x_o, y_o)dx_o\,dy_o \tag{4.34}$$

$h(x_i - mx_o, y_i - my_o)$ is termed the impulse response of the imaging system since it gives the response of an impulse in spatial coordinates, i.e. a delta function. For an ideal lens the

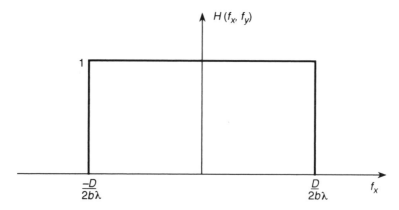

Fig. 4.20.   *The coherent transfer function for a diffraction-limited system*

impulse response is given by Equation (4.27). Such a lens is said to be diffraction limited since its performance is limited only by the size of the exit pupil.

Equation (4.34) is seen to be a convolution integral (see Equation (B.2e) in Appendix B.1). According to the convolution theorem (see Appendix B.3) this implies that

$$G_i(f_x, f_y) = H(f_x, f_y)G_o(f_x, f_y) \tag{4.35}$$

where G_i, G_o and H are the Fourier transforms of u_i, u_o and h respectively. From this expression we see that the spatial frequency distribution $G_i(f_x, f_y)$ of the image is equal to the spatial frequency distribution $G_o(f_x, f_y)$ of the object times the function $H(f_x, f_y)$ which is referred to as the coherent transfer function. $H(f_x, f_y)$ acts as a filter, filtering out the spatial frequencies (the plane wave components) not accepted by the lens aperture. We have that, since

$$\mathscr{F}\mathscr{F}\{g(x, y)\} = g(-x, -y) \tag{4.36}$$

$$H(f_x, f_y) = \mathscr{F}\{h(x_i, y_i)\} = \mathscr{F}\mathscr{F}\{P(x, y)\} = P(-\lambda bf_x, -\lambda bf_y) \tag{4.37}$$

For a circularly symmetric aperture function we have $P(-\lambda bf_x, -\lambda bf_y) = P(\lambda bf_x, \lambda bf_y)$. Figure 4.20 shows a plot of the coherent transfer function for a diffraction limited lens with a circular aperture along the f_x-axis. We see that the spatial frequencies f_x are unaffected up to f_o, the maximum spatial frequency resolved by the lens given by

$$\lambda bf_o = \frac{D}{2}, \qquad \text{that is} \qquad f_o = \frac{D}{2b\lambda} \tag{4.38}$$

which is called the cut-off frequency of the coherent transfer function.

4.6.2 The incoherent transfer function

If the object plane contains a collection of point sources which are mutually incoherent, we get the light distribution in the image plane not by adding the field amplitudes, but by adding the intensities contributed by each point. In accordance with Equation (4.34) we then get for the intensity

$$I_i(x_i, y_i) = \int\int_{-\infty}^{\infty} |h(x_i - mx_o, y_i - my_o)|^2 I_o(x_o, y_o) dx_o dy_o \tag{4.39}$$

In analogy with Equation (4.37) we therefore can introduce an incoherent transfer function $\mathscr{H}(f_x, f_y)$ given by

$$\mathscr{H}(f_x, f_y) = \frac{\int\int_{-\infty}^{\infty} |h(x_i, y_i)|^2 e^{-i2\pi(f_x x_i + f_y y_i)} dx_i dy_i}{\int\int_{-\infty}^{\infty} |h(x_i, y_i)|^2 dx_i dy_i} \tag{4.40}$$

Where we by the denominator have normalized the transfer function, so that $\mathscr{H}(0,0) = 1$.

The relationship between the coherent and the incoherent transfer function is therefore given by

$$H(f_x, f_y) = \mathscr{F}\{h\} \tag{4.41a}$$

$$\mathscr{H}(f_x, f_y) = \frac{\mathscr{F}\{|h|^2\}}{\mathscr{F}\{|h|^2\}_{f_x = f_y = 0}} \tag{4.41b}$$

From the autocorrelation theorem (see Equation (B.2f), Appendix B.1) it follows that

$$\mathscr{H}(f_x, f_y) = \frac{\int\int_{-\infty}^{\infty} H(\xi', \eta') H^*(\xi' + f_x, \eta' + f_y) d\xi' d\eta'}{\int\int_{-\infty}^{\infty} |H(\xi', \eta')|^2 d\xi' d\eta'} \tag{4.42}$$

A simple change of variables

$$\xi = \xi' + \frac{f_x}{2} \tag{4.43a}$$

$$\eta = \eta' + \frac{f_y}{2} \tag{4.43b}$$

results in the symmetrical expression

$$\mathscr{H}(f_x, f_y) = \frac{\int\int_{-\infty}^{\infty} H\left(\xi - \frac{f_x}{2}, \eta - \frac{f_y}{2}\right) H^*\left(\xi + \frac{f_x}{2}, \eta + \frac{f_y}{2}\right) d\xi d\eta}{\int\int_{-\infty}^{\infty} |H(\xi, \eta)|^2 d\xi d\eta} \tag{4.44}$$

To this point, our discussion has been equally applicable to systems with and without abberations. We now consider the special case of a diffraction-limited incoherent system. Recall that for the coherent system we have, cf. Equation (4.37)

$$H(f_x, f_y) = P(\lambda b f_x, \lambda b f_y) \tag{4.45}$$

For the incoherent system it follows from Equation (4.44) (with a simple change of variables) that

$$\mathscr{H}(f_x, f_y) = \frac{\displaystyle\iint_{-\infty}^{\infty} P\!\left(\xi - \frac{\lambda b f_x}{2}, \eta - \frac{\lambda b f_y}{2}\right) P\!\left(\xi + \frac{\lambda b f_x}{2}, \eta + \frac{\lambda b f_y}{2}\right) d\xi\, d\eta}{\displaystyle\iint_{-\infty}^{\infty} |P(\xi, \eta)|\, d\xi\, d\eta} \tag{4.46}$$

where in the denominator, since $|P|$ equals either unity or zero, $|P|^2$ has been replaced by $|P|$.

The geometrical interpretation of Equation (4.46) is that the numerator represents the area of overlap of two displaced pupil functions, one centred at $(\lambda b f_x/2, \lambda b f_y/2)$ and the second centred at $(-\lambda b f_x/2, -\lambda b f_y/2)$. The denominator simply normalizes this area of overlap by the total area of the pupil. Note that this interpretation demonstrates that $\mathscr{H}(f_x, f_y)$ of a diffraction-limited system is always real and non-negative.

As an example, consider Figure 4.21 where the pupil is a square of width l. The area of overlap is

$$A(f_x, f_y) = \begin{cases} (1 - \lambda b|f_x|)(1 - \lambda b|f_y|) & |f_x| \le \dfrac{1}{\lambda b} \quad |f_y| \le \dfrac{1}{\lambda b} \\ 0 & \text{otherwise} \end{cases} \tag{4.47}$$

When dividing by the total area l^2, the result becomes

$$\mathscr{H}(f_x, f_y) = \Lambda\!\left(\frac{\lambda b f_x}{l}\right)\Lambda\!\left(\frac{\lambda b f_y}{l}\right) = \Lambda\!\left(\frac{f_x}{2f_0}\right)\Lambda\!\left(\frac{f_y}{2f_0}\right) \tag{4.48}$$

where Λ is the triangle function (see Equation (B.15), Appendix B.2) and f_0 is the cut-off frequency of the same system used with coherent illumination

$$f_0 = \frac{1}{2\lambda b} \tag{4.49}$$

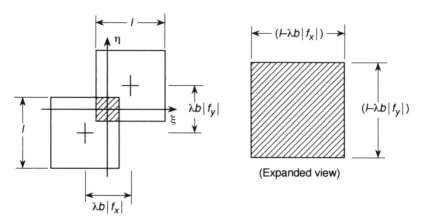

Fig. 4.21. *Geometry of the area of overlap when calculating the optical transfer function of a diffraction-limited system with a square pupil*

We therefore have that the cut-off frequency for the incoherent transfer function is twice that of the coherent transfer function.

$\mathscr{H}(f_x, f_y)$ is commonly known as the optical transfer function (OTF) of the system. Generally it is a complex quantity and its modulus $|\mathscr{H}|$ is known as the modulation transfer function (MTF). When abberations are present, this can be accounted for by introducing a generalized pupil function

$$P(x, y) = |P(x, y)| e^{ikW(x,y)} \tag{4.50}$$

where $k = 2\pi/\lambda$. Here $P(x, y)$ represents a phase-shifting plate while $W(x, y)$ is an effective path-length error deforming the ideal, converging spherical wavefront representing the diffraction-limited case.

Since imaging systems are usually symmetric about the optical axis it is sufficient to display the MTF for positive frequencies along one axis (the f_x-axis). Figure 4.22 shows $|\mathscr{H}(f_x)|$ from Equation (4.48) representing the MTF for a diffraction-limited system (curve (a)). Curve (b)) shows the MTF for a 'soft' lens system. The term 'soft' comes from the fact that this lens will enhance low frequency components relative to high-frequency components. The image in such a system will appear soft. Curve (c) shows the MTF for a 'hard' lens system.

The MTF-concept is adopted from electrical engineering. It means that if we have an 'input' object with an intensity distribution

$$I_o = a(1 + b \cos 2\pi f_g x) \tag{4.51}$$

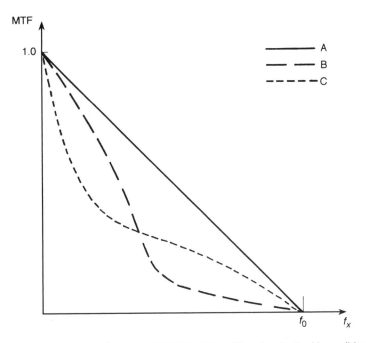

Fig. 4.22. *Modulation transfer function (MTF) for (a) a diffraction limited lens, (b) a 'soft' lens, and (c) a 'hard' lens*

Its spectrum becomes

$$J_o = \mathscr{F}\{I_o\} = a\delta(f_x) + \frac{ab}{2}\delta(f_x - f_g) + \frac{ab}{2}\delta(f_x + f_g) \qquad (4.52)$$

The spectrum of the 'output' image becomes

$$\mathscr{F}\{I_i\} = J_i = a\mathscr{H}\delta(f_x) + \frac{ab}{2}\mathscr{H}\delta(f_x - f_g) + \mathscr{H}\frac{ab}{2}\delta(f_x + f_g) \qquad (4.53)$$

which by taking the inverse transform gives the 'output' image

$$I_i = a[1 + |\mathscr{H}(f_g)|b\cos(2\pi f_g x)] \qquad (4.54)$$

This means that the modulation (or the visibility, see Section 3.2) of the input sinusoidal grating has changed by a factor $|\mathscr{H}(f_g)|$, the value of the MTF at frequency f_g.

When comparing coherent and incoherent imaging, we see that the image intensity in the two cases are given by

$$\text{Incoheren:} \quad I_i = |h|^2 \otimes I_o = |h|^2 \otimes |u_o|^2 \qquad (4.55a)$$

$$\text{Coherent:} \quad I_i = |h \otimes u_o|^2 \qquad (4.55b)$$

From the definition of the autocorrelation integral (see Equation (B.2f), Appendix B.1) we therefore can write the frequency spectra of the image intensity in the two cases as

$$\text{Incoherent:} \quad \mathscr{F}\{I_i\} = [H \odot H][G_o \odot G_o] \qquad (4.56a)$$

$$\text{Coherent:} \quad \mathscr{F}\{I_i\} = HG_o \odot HG_o \qquad (4.56b)$$

4.6.3 The depth of focus

Another aspect of the imaging properties of a lens is the depth of focus. By this is meant the maximum distance by which the image plane can be moved away from the exact focus and still have an acceptable focused image. Consider Figure 4.23 where the rays passing through the edge of the lens aperture from an on-axis point source are drawn. The image plane is moved a distance Δb away from the exact image point. Let us take the depth of focus to be the distance Δb when the radius of the light spot on the image plane is equal to the radius of the Airy disc Δr_f. We then have

$$\frac{\Delta r_f}{\Delta b} = \frac{D/2}{b} \qquad (4.57)$$

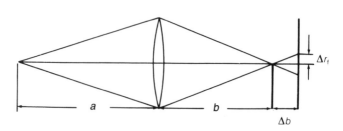

Fig. 4.23

which by inserting the value of the radius of the Airy disc, Equation (4.31), gives

$$\Delta b = 2.44\lambda\left(\frac{b}{D}\right)^2 = 2.44\lambda\left(\frac{bF}{f}\right)^2 = 2.44\lambda(1+m)^2F^2 \tag{4.58}$$

Here $F = f/D$ is the aperture number, m is the magnification and f is the focal length. From this expression we see that the depth of focus is inversely proportional to the square of the lens aperture or proportional to the square of the aperture number.

4.7   THE PHASE-MODULATED SINUSOIDAL GRATING

Consider a sinusoidal grating given by

$$t_1 = a(1 + \cos\phi_1) = a\left(1 + \cos\frac{2\pi x}{d}\right) \tag{4.59}$$

Assume that this grating undergoes a change whereafter it is given by

$$t_2 = a(1 + \cos\phi_2) = a\left[1 + \cos\left(\frac{2\pi x}{d} + \psi(x)\right)\right] \tag{4.60}$$

This change might result from different causes, which will be taken up later. We assume that $\psi(x)$ varies much more slowly than $2\pi x/d$. The spectrum of such a grating is indicated in Figure 4.24(b) which should be compared with the spectrum of the grating given in Equation (4.59) in Figure 4.24(a). We see that the side orders have become broader because of the slowly varying function $\psi(x)$. The grating is said to be phase-modulated and $\psi(x)$ is called the modulation function.

We then image these gratings onto a photographic film resulting in an intensity distribution (blackening) on the film given by

$$I(x) = t_1 + t_2 = a(1 + \cos\phi_1) + a(1 + \cos\phi_2)$$

$$= 2a\left(1 + \cos\frac{\phi_1 - \phi_2}{2}\cos\frac{\phi_1 + \phi_2}{2}\right) \tag{4.61}$$

Fig. 4.24. *The spectrum of (a) a sinusoidal grating of period d, (b) the same grating phase-modulated*

After development, $I(x)$ becomes equal to the transmittance function of the film. We then place this film in the object plane in the set-up of Figure 4.12. The intensity distribution in the image plane then becomes equal to Equation (4.61) with a scaling factor determined by the magnification m. Equation (4.61) can be written as

$$I(x) = 2a + \frac{a}{2}(e^{i\phi_1} + e^{i\phi_2}) + \frac{a}{2}(e^{-i\phi_1} + e^{-i\phi_2}) \tag{4.62}$$

where the first term represents the zeroth order and the second and third terms represent the two side orders. In the filter plane we place a screen with a hole a distance $\lambda f/d$ from the optical axis and with an opening wide enough to transmit the full width of one of the modulated side orders given in Figure 4.24(b). The wave field in the image plane then becomes (assuming a magnification $m = 1$)

$$u = \frac{a}{2}(e^{i\phi_1} + e^{i\phi_2}) \tag{4.63}$$

and the intensity

$$I = |u|^2 = \frac{a^2}{2}[1 + \cos(\phi_1 - \phi_2)] = \frac{a^2}{2}[1 + \cos\psi(x)] \tag{4.64}$$

From this equation we see that we have obtained an expression dependent on the modulation function alone. The grating is said to be demodulated.

Another procedure is to image the two gratings given in Equations (4.59) and (4.60) onto two separate negatives and then put them together in the image plane. The resulting transmittance function then becomes not the sum but the product

$$\begin{aligned}
t(x) = t_1 t_2 &= a(1 + \cos\phi_1)a(1 + \cos\phi_2) \\
&= a^2(1 + \cos\phi_1 + \cos\phi_2 + \cos\phi_1\cos\phi_2) \\
&= a^2[1 + \cos\phi_1 + \cos\phi_2 + \tfrac{1}{2}\cos(\phi_1 - \phi_2) + \tfrac{1}{2}\cos(\phi_1 + \phi_2)] \\
&= a^2\left[1 + \tfrac{1}{2}\cos\psi(x) + \cos\frac{2\pi x}{d} + \cos\left(\frac{2\pi x}{d} + \psi(x)\right)\right. \\
&\quad \left. + \tfrac{1}{2}\cos\left(\frac{4\pi x}{d} + \psi(x)\right)\right]
\end{aligned} \tag{4.65}$$

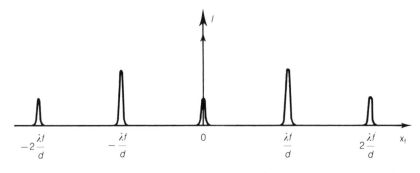

Fig. 4.25. *The spectrum of the gratings in Figure 4.24 (a, b) multiplied*

The spectrum of this function is sketched in Figure 4.25. The second term in Equation (4.65) gives the modulation function directly, i.e. it modulates the zeroth order. In practice, the zeroth order is often much stronger than the modulation function, resulting in a pattern of very low contrast. We then have the possibility of filtering out the first side order and obtaining the same result as in the first procedure.

PROBLEMS

4.1 Suppose that we have a laser emitting a diffraction-limited beam ($\lambda_o = 632.8$ nm) having a 2 mm diameter. How big a light spot would be produced on the surface of the moon a distance of 376×10^3 km away? Neglect any effects of the Earth's atmosphere.

4.2 Imagine that you are looking through a piece of square woven cloth at a point source ($\lambda_o = 600$ nm) 20 m away. If you see a square arrangement of bright spots located about the point source, each separated by an apparent nearest-neighbour distance of 12 cm, how close together are the strands of cloth?

4.3 Imagine an opaque screen containing thirty randomly located circular holes of the same diameter. The light source is such that every aperture is coherently illuminated by its own plane wave. Each wave in turn is completely incoherent with respect to all the others. Describe the resulting far-field diffraction pattern.

4.4 Determine the Fourier transform of the function

$$U(x) = \begin{cases} U_0 \sin 2\pi f_0 x & |x| \leq L \\ 0 & |x| > L \end{cases}$$

Make a sketch of it.

4.5 Determine the Fourier transform of

$$f(x) = \begin{cases} \sin^2 2\pi f_0 x & |x| \leq L \\ 0 & |x| > L \end{cases}$$

Make a sketch of it.

4.6 Suppose we have two functions, $f(x, y)$ and $h(x, y)$, where both have a value of 1 over a square region in the xy-plane and are zero everywhere else (Figure P4.1). Given that $g(x, y)$ is their convolution, make a plot of $g(x, 0)$.

4.7 Calculate and sketch the convolution between the two functions $f(x)$ and $h(x)$ depicted in Figure P4.2.

4.8 Make a sketch of the resulting function arising from the convolution of the two functions depicted in Figure P4.3.

4.9 Show (for normally incident plane waves) that if an aperture has a centre of symmetry, i.e. if the aperture function is even, then the diffracted field in the Fraunhofer case also possesses a centre of symmetry.

Fig. P4.1

Fig. P4.2

Fig. P4.3

4.10 Another way to do the integral for Fraunhofer diffraction by a circular aperture is to stay in Cartesian coordinates. If the Fourier transform of $f(x, y)$ is $F(f_x, f_y)$ it is sufficient to calculate $F(f_x, 0)$ or $F(0, f_y)$ since we know that $F(f_x, f_y)$ also is circularly symmetric. In this way, calculate the Fourier transform of a circular aperture and show that

$$F(\rho) = 4r_0^2 \int_0^1 \sqrt{(1 - u^2)} \cos(tu)\, \mathrm{d}u$$

where $t = 2\pi\rho r_0$ and r_0 is the radius of the aperture. By comparison with Equation (4.27) we obtain another representation of the Bessel function $J_1(t)$:

$$J_1(t) = \frac{2t}{\pi} \int_0^1 \sqrt{(1 - u^2)} \cos(tu)\, du$$

4.11 A real function $f(x)$ can be decomposed into a sum of an even and an odd function.

(a) Show that $f_{even}(x) = \frac{1}{2}[f(x) + f(-x)]$ and $f_{odd}(x) = \frac{1}{2}[f(x) - f(-x)]$.
(b) Show that

$$\mathscr{F}\{f_{even}(x)\} = \text{Re}\{\mathscr{F}\{f(x)\}\}$$
$$\mathscr{F}\{f_{odd}(x)\} = i\,\text{Im}\{\mathscr{F}\{f(x)\}\}$$

4.12 The sign-function is defined as

$$\text{sgn}(x) = \begin{cases} 1 & x > 0 \\ 0 & x = 0 \\ -1 & x < 0 \end{cases}$$

Prove that

$$\mathscr{F}\{\text{sgn}(x)\,\text{sgn}(y)\} = \left(\frac{1}{i\pi f_x}\right)\left(\frac{1}{i\pi f_y}\right)$$

4.13 Prove that

$$\mathscr{F}\{\nabla^2 g(x, y)\} = -4\pi^2(f_x^2 + f_y^2)\mathscr{F}\{g(x, y)\}$$

where ∇^2 is the Laplacian operator

$$\nabla^2 = \frac{\partial^2}{\partial x^2} + \frac{\partial^2}{\partial y^2}$$

4.14 Prove that

$$\mathscr{F}\mathscr{F}\{g(x, y)\} = \mathscr{F}^{-1}\mathscr{F}^{-1}\{g(x, y)\} = g(-x, -y)$$

4.15 The Gaussian distribution function is given as

$$g(x) = C e^{-ax^2}$$

where

$$C = \sqrt{\left(\frac{a}{x}\right)}, \qquad \sigma_x = \frac{1}{\sqrt{(2a)}}$$

where σ_x is the standard deviation. Given that

$$\int_{-\infty}^{\infty} e^{-u^2}\, du = \sqrt{\pi}$$

prove that the Fourier transform of $g(x)$ is

$$G(f_x) = \exp\left[-\frac{(2\pi f_x)^2}{4a}\right]$$

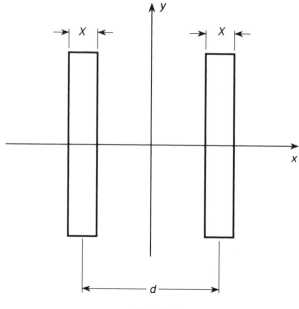

Fig. P4.4

4.16 Assuming a unit-amplitude, normally incident plane-wave illumination:

(a) Find the intensity distribution in the Fraunhofer diffraction pattern of the double-slit aperture shown in Figure P4.4.

(b) Sketch the intensity distribution along the x_f-axis of the observation plane. Let $X/\lambda z = 1\,\mathrm{m}^{-1}$, and $d/\lambda z = 3/2\,\mathrm{m}^{-1}$, where z is the distance to the observation plane and λ the wavelength.

4.17 (a) Sketch the aperture described by the transmittance function

$$t(x,y) = \left\{ \left[\mathrm{rect}\left(\frac{x}{X}\right)\mathrm{rect}\left(\frac{y}{Y}\right) \right] \otimes \left[\frac{1}{d}\mathrm{comb}\left(\frac{x}{d}\right)\delta(y) \right] \right\} \mathrm{rect}\left(\frac{x}{Nd}\right)$$

where N is an odd integer and $d > X$.

(b) Find an expression for the intensity distribution in the Fraunhofer diffraction pattern of that aperture, assuming illumination by a normally incident plane wave and $N \gg 1$.

4.18 Find an expression for the intensity distribution in the Fraunhofer diffraction pattern of the aperture shown in Figure P4.5. Assume unit-amplitude, normally incident plane-wave illumination.

4.19 A normally incident, unit-amplitude, monochromatic ($\lambda = 1\,\mu\mathrm{m}$) plane wave illuminates a positive lens of 40 mm diameter and 2 m of focal length. An object is placed 1 m behind the lens and centred on the lens axis with amplitude transmittance

$$t(x,y) = \tfrac{1}{2}(1 + \cos 2\pi f_0 x)\,\mathrm{rect}\left(\frac{x}{L}\right)\mathrm{rect}\left(\frac{y}{L}\right)$$

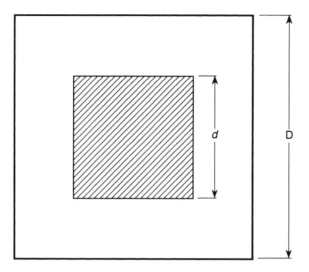

Fig. P4.5

Assuming $L = 10\,\text{mm}$ and $f_0 = 10\,\text{lines/mm}$, sketch the intensity distribution across the x_f-axis of the focal plane. Indicate the numerical values of the distance between the diffracted components and the width (between first zeros) of the individual components.

4.20 An incoherent imaging system has a square pupil function of width D. A square stop, of width $D/2$, is placed at the centre of the exit pupil, as shown in Figure P4.5.

(a) Sketch a cross-section of the optical transfer function $\mathscr{H}(f_x, 0)$ along the f_x-axis with and without the stop present.
(b) Sketch the limiting form of the optical transfer function as the size of the stop approaches the size of the exit pupil.

4.21 A sinusoidal amplitude grating with amplitude transmittance

$$t(x_o, y_o) = \tfrac{1}{2} + \tfrac{1}{2}\cos 2\pi f_1 x_o$$

is placed in the object plane x_o, y_o and imaged to the image plane x_i, y_i by a lens (circular of diameter D, focal length f) and obliquely illuminated by a unit-amplitude monochromatic plane wave incident at an angle θ in the $x_o z$-plane. The object- and image distances are a and b respectively.

(a) Find the Fourier spectrum of the field-amplitude distribution transmitted by the object screen.
(b) Assuming $a = b = 2f$, what is the maximum angle θ_m for which any variations of intensity will appear in the image plane?
(c) By applying this maximum angle θ_m, what is the intensity distribution in the image plane, and how does it compare with the corresponding intensity distribution for $\theta = 0$?
(d) Assuming that the maximum angle θ_m is used. Find the maximum grating frequency f_1 that will give any variations of intensity in the image plane. Compare this frequency with the cutoff frequency when $\theta = 0$.

Fig. P4.6

4.22 An object has an intensity transmittance given by

$$\tau(x, y) = \tfrac{1}{2}(1 + \cos 2\pi f_1 x)$$

and introduces a constant, uniform phase delay across the object plane. This object is imaged by a positive lens of diameter D and focal length f, with an object and image distance $a = b = 2f$. Compare the maximum frequencies f_1 transmitted by the system for the case of coherent and incoherent illumination.

4.23 An object has a periodic amplitude transmittance described by

$$t(x, y) = t(x)$$

where $t(x)$ is shown in Figure P4.6. This object is placed in the object plane of a lens with object- and image distance $a = b = 2f$. A small opaque stop is introduced on the lens axis in the focal plane. Sketch the resulting intensity distribution in the image plane.

4.24 The so-called central dark ground method for observing phase objects is achieved by placing a small opaque stop in the back focal plane of the imaging lens to block the undiffracted light. Assuming that the phase delay $\phi(x, y)$ through the object is always much less than 1 radian, find the observed image intensity in terms of $\phi(x, y)$.

4.25 According to the so-called Rayleigh criterion of resolution, two incoherent point sources are just resolved by a diffraction-limited system when the centre of the Airy pattern generated by one source falls on the first zero of the Airy pattern generated by the second. Assume that the image points of two points just resolved are centred at $(s/2, 0)$ and $(-s/2, 0)$ in the image plane, where s is the minimum resolvable separation.

(a) Find s in terms of the exit pupil diameter D, the image distance b and the wavelength. The intensity distribution along the x-axis in the image plane will consist of two maxima on each side of a central dip.

(b) Calculate the ratio between the intensity of the central dip and the maximum. $J_1(0.61\pi) = 0.58$.

5 Light sources and detectors

5.1 RADIOMETRY. PHOTOMETRY

To compare light sources we have to make a brief introduction to units and terms for the measurement of electromagnetic radiation (Slater 1980; Klein 1970; Longhurst 1967). Below we present the most common radiometric units.

Radiant energy, Q, is energy travelling in the form of electromagnetic waves, measured in joules.

Radiant flux, $\Phi = \partial Q/\partial t$ is the time rate of change, or rate of transfer, of radiant energy, measured in watts. Power is equivalent to, and often used instead of, flux. Radiant flux density at a surface, $M = E = \partial \Phi/\partial A$, is the radiant flux at a surface divided by the area of the surface. When referring to the radiant flux emitted from a surface it is called radiant exitance M. When referring to the radiant flux incident on a surface it is called irradiance E. Both are measured in watts per square metre. Note that in the rest of this book, we use the term intensity, which is proportional to irradiance.

Radiant intensity, $I = \partial \Phi/\partial \Omega$, of a source is the radiant flux proceeding from the source per unit solid angle in the direction considered, measured in watts per steradian.

Radiance, $L = \partial^2 \Phi/\partial \Omega \, \partial A \cos \theta$, in a given direction, is the radiant flux leaving an element of a surface and propagated in directions defined by an elementary cone containing the given direction, divided by the product of the solid angle of the cone and the area of the projection of the surface element on a plane perpendicular to the given direction. Figure 5.1 illustrates the concept of radiance. It is measured in watts per square metre and steradian.

All of the radiometric terms have their photometric counterparts. They are related to how the (standard) human eye respond to optical radiation and is limited to the visible part of the spectrum. In Table 5.1 we list the radiometric and the corresponding photometric quantities.

To distinguish radiometric and photometric symbols they are given subscripts e and v respectively (e.g. L_e = radiance, L_v = luminance).

The radiometric quantities refer to total radiation of all wavelengths. A spectral version for each may be defined by adding the subscript λ (e.g. $M_{e\lambda}$ or simply M_λ) where for example a spectral flux $\Phi_\lambda \, d\lambda$ represents the flux in a wavelength interval between λ and $\lambda + d\lambda$, with units watts per nanometre (W nm^{-1}) or watt per micrometre $(\text{W}\,\mu\text{m}^{-1})$.

To represent the response of the human eye, a standard luminosity curve $V(\lambda)$ has been established, see Figure 5.2. It has a peak value of unity at $\lambda = 555$ nm. The conversion at $\lambda = 555$ nm is standardized to be

$$K_m = 680 \, \text{lm/W} \tag{5.1}$$

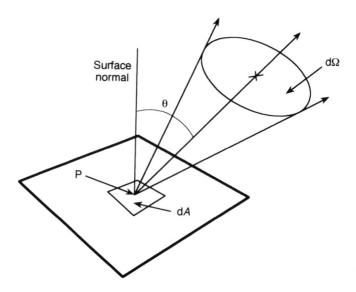

Fig. 5.1. *The concept of radiance*

Table 5.1 *Symbols, standard units and defining equations for fundamental radiometric and photometric quantities*

Symbol	Radiometric quantity	Radiometric units	Defining equation	Photometric quantity	Photometric units
Q	Radiant energy	J		Luminous energy	lm s
Φ	Radiant flux	W	$\Phi = \partial Q/\partial t$	Luminos flux	lm
M	Radiant exitance	$W\,m^{-2}$	$M = \partial\Phi/\partial A$	Luminous exitance	$lm\,m^{-2}$
E	Irradiance	$W\,m^{-2}$	$E = \partial\Phi/\partial A$	Illuminance	$lm\,m^{-2}$
I	Radiant intensity	$W\,sr^{-1}$	$I = \partial\Phi/\partial\Omega$	Luminous intensity	$lm\,sr^{-1}$
L	Radiance	$W\,sr^{-1}\,m^{-2}$	$L = \partial^2\Phi/\partial\Omega\partial A\cos\theta$	Luminance	$lm\,sr^{-1}\,m^{-2}$

This means that 1 W of flux at 555 nm gives the same physical sensation as 680 lm. For other wavelengths the conversion factor is

$$K = K_m V(\lambda) = 680 V(\lambda)\,lm/W \tag{5.2}$$

We may use Equation (5.2) to convert any radiometric quantity to the corresponding photometric quantity. For instance, if we have a spectral radiant flux $\Phi_{e\lambda}$, the luminous

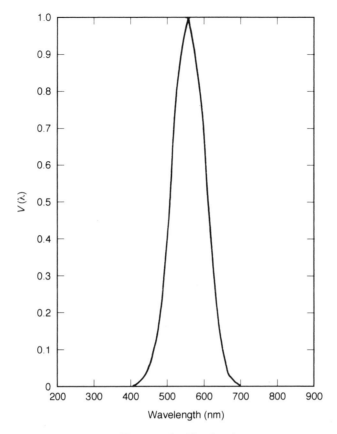

Fig. 5.2. *The standard luminosity curve*

flux is given by

$$\Phi_v = 680 \int_0^\infty V(\lambda)\Phi_{e\lambda}\,d\lambda \qquad (5.3)$$

An average conversion factor is defined by

$$K_{av} = \Phi_v/\Phi_e \qquad (5.4)$$

5.1.1 Lambertian surface

A Lambertian surface is a perfectly diffuse reflecting surface defined as one which the radiance L is constant for any angle of reflection θ to the surface normal. Lambert's cosine law states that the intensity (flux per unit solid angle) in any direction varies as the cosine of the reflection angle:

$$I = I_0 \cos\theta \qquad (5.5)$$

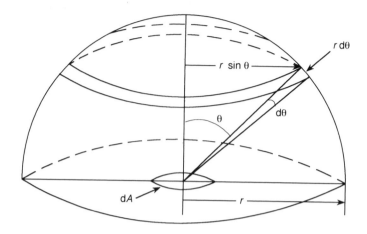

Fig. 5.3

Since the projected area of the source also varies as $\cos\theta$, the radiance becomes independent of the viewing angle:

$$L = \frac{I}{dA\cos\theta} = \frac{I_0}{dA} \tag{5.6}$$

Assume that an elemental Lambertian surface dA is irradiated by E in W m^{-2} and that the radiant flux reflected in any direction θ to the surface normal is given by the basic equation

$$d^2\Phi = L\,d\Omega\,dA\cos\theta \tag{5.7}$$

The solid angle $d\Omega$ in polar coordinates (see Figure 5.3) is given by

$$d\Omega = (r\sin\theta\,d\theta r\,d\phi)/r^2 = \sin\theta\,d\theta\,d\phi \tag{5.8}$$

The total radiant flux reflected into the hemisphere therefore is given by

$$d\Phi_h = \int_0^{2\pi} d\varphi \int_0^{\pi/2} L\,dA\cos\theta\sin\theta\,d\theta = \pi L\,dA \tag{5.9}$$

The ratio of the total reflected radiant flux to the incident radiant flux $d\Phi_i = E\,dA$ defines the diffuse reflectance of the surface

$$\frac{d\Phi_h}{d\Phi_i} = \rho = \frac{\pi L}{E} \tag{5.10}$$

The quantity ρE is the radiant flux density reflected from the surface which is equivalent to the radiant exitance M of a self-emitting source, giving

$$M = \pi L \tag{5.11}$$

for a Lambertian surface.

For non-Lambertian surfaces, L is a function of both θ and the azimuthal angle ϕ and therefore can not be taken outside the integral in Equation (5.9). Many natural surfaces show Lambertian characteristics up to $\theta = 40°$. In satellite observations, one has found snow and desert to be Lambertian up to about 50° or 60°. Most naturally occurring surfaces depart significantly from the Lambertian case for θ greater than about 60°, an exception is White Sands, the desert in New Mexico, which is nearly Lambertian for all angles.

5.1.2 Blackbody radiator

A blackbody at a given temperature provides the maximum radiant exitance at any wavelength that any body in thermodynamic equilibrium at that temperature can provide. It follows that a blackbody is a Lambertian source and that it is a perfect absorber as well as a perfect radiator. The spectral radiant exitance M_λ from a blackbody is given by Planck's formula

$$M_\lambda = \frac{2\pi hc^2}{\lambda^5[\exp(hc/\lambda kT) - 1]}$$ (5.12)

where

    h = Planck's constant = 6.6256×10^{-34} J s;
    c = velocity of light = 2.997925×10^8 ms^{-1};
    k = Boltzmann's constant = 1.38054×10^{-23} J K^{-1};
    T = absolute temperature in kelvin;
    λ = wavelength in metres.

which gives M_λ in W m^{-2} µm^{-1}. Figure 5.4 shows M_λ as a function of wavelength for different temperatures.

Fig. 5.4. *Spectral radiant exitance from a blackbody at various temperatures according to Planck's law*

By integrating over all wavelengths we get the Stefan–Boltzmann law

$$M = \int_0^\infty M_\lambda \, d\lambda = \sigma T^4 \tag{5.13}$$

where $\sigma = (2\pi^5 k^4)/(15c^2 h^3) = 5.672 \times 10^{-8} \, \mathrm{W\,m^{-2}\,K^{-4}}$ is called the Stefan–Boltzmann constant.

By differentiating Equation (5.12) we get the wavelength λ_m for which M_λ is peaked

$$\lambda_m T = 2897.8 \, \mu\mathrm{m\,K} \tag{5.14}$$

This relation is called Wien's displacement law.

The blackbody is an idealization. In nature most radiators are selective radiators, i.e. the spectral distribution of the emitted flux is not the same as for a blackbody. Emissitivity is a measure of how a real source compares with a blackbody and is defined as

$$\varepsilon = M'/M \tag{5.15}$$

where M' is the radiant exitance of the source of interest and M is the radiant exitance of a blackbody at the same temperature. ε is a number between 0 and 1 and is in general both wavelength and temperature dependent. When ε is independent of wavelength the source is called a greybody. A more general form of Equation (5.15) can be written to take into account the spectrally varying quantities, thus ε, the emissivity for a selective radiator, as an average over all wavelengths is

$$\varepsilon = \frac{M'}{M} = \frac{\displaystyle\int_0^\infty \varepsilon(\lambda) M_\lambda \, d\lambda}{\displaystyle\int_0^\infty M_\lambda \, d\lambda} = \frac{1}{\sigma T^4} \int_0^\infty \varepsilon(\lambda) M_\lambda \, d\lambda \tag{5.16}$$

Consider two slabs of different materials A and B, and that each is of semi-infinite thickness and infinite area, forming a cavity as shown in Figure 5.5. Assume that A is a blackbody and that B is a material with emissivity ε, reflectance ρ and absorbtance α, and that the materials and cavity are in thermal equilibrium. Because of the last assumption, the flux onto B must equal the flux leaving B toward A. Thus

$$M_{bb} = \rho M_{bb} + M \tag{5.17}$$

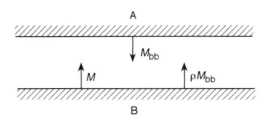

Fig. 5.5.   *Radiant exitances between a blackbody A and another material B*

where M_{bb} and M are the radiant exitance of A (the blackbody) and B respectively. From the definition of emissivity we have

$$M/M_{bb} = \varepsilon = 1 - \rho \tag{5.18}$$

which is referred to as Kirchhoff's law.

Because of conservation of energy, the reflectance, transmittance and absorbance at a surface add up to unity. Since we have assumed semi-infinitely thick materials, the transmittance is zero and we have

$$M/M_{bb} = \varepsilon = 1 - \rho = \alpha \tag{5.19}$$

where α is the absorptance of material B. Equation (5.19) states that good emitters and absorbers are poor reflectors and vice versa. We can anticipate that Equation (5.19) holds for any given spectral interval which gives the more general form

$$M_{\lambda}/M_{\lambda bb} = \varepsilon(\lambda) = 1 - \rho(\lambda) = \alpha(\lambda) \tag{5.20}$$

5.1.3 Examples

Let us compare the light from a typical He–Ne laser and a blackbody with the same area as the output aperture of the laser. Assume this area to be $1\,\text{mm}^2$ and the blackbody temperature to be 3000 K, close to the temperature of the filament of an incandescent lamp. From Equation (5.13) we find the blackbody exitance to be $4.6 \times 10^6\,\text{W}\,\text{m}^{-2}$ which gives a radiant flux of 4.6 W. An ordinary He–Ne laser has an output of about 1/1000 th of this, not very impressive even if we take into account that most of the radiation from the blackbody is outside the visible region.

From Equation (5.11) we find the radiance from the blackbody to be

$$L = M/\pi = 1.46 \times 10^6\,\text{W}\,\text{m}^{-2}\,\text{sr}^{-1}$$

The light beam from the laser has a diverging angle of about λ/d where λ is the wavelength and d is the output aperture diameter. This gives a solid angle of about λ^2/A where A is the aperture area. The radiance at the centre of the beam is therefore ($\cos\theta = 0$)

$$L = \frac{\Phi}{\Omega A} = \frac{\Phi}{\lambda^2} \tag{5.21}$$

With a radiant power (flux) $\Phi = 5\,\text{mW}$ and a wavelength $\lambda = 0.6328\,\mu\text{m}$, this gives $L = 1.2 \times 10^{10}\,\text{W}\,\text{m}^{-2}\,\text{sr}^{-1}$, a number clearly in favour of the laser. Note that the radiance of the blackbody is independent of its area. By decreasing the power of the laser by reducing its output aperture, the radiance decreases accordingly.

Figure 5.6 illustrates the imaging of an object of elemental area dA_o by a lens system with the entrance and exit pupils as sketched. We assume that the object is a Lambertian surface of radiance L_o. The flux incident over an annular element of the entrance pupil is given by

$$d^2\Phi = L_o\,dA_o\cos\theta\,d\Omega \tag{5.22}$$

where

$$d\Omega = 2\pi\sin\theta\,d\theta \tag{5.23}$$

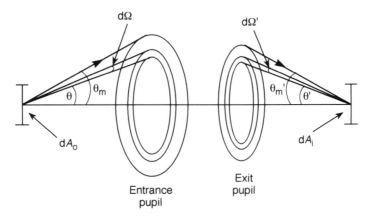

Fig. 5.6. *Geometry for determining the radiometry of an optical system*

If θ_m is the angle of the marginal ray passing through the entrance pupil, the flux incident over the entrance pupil is

$$d\Phi_o = 2\pi L_o \, dA_o \int_0^{\theta_m} \sin\theta \cos\theta \, d\theta = \pi L_o \, dA_o \sin^2\theta_m \tag{5.24}$$

Equation (5.24) is not the product of radiance, area and solid angle, or $2\pi L_o dA_o(1 - \cos\theta_m)$, as we might at first expect, because the cosine factor, which accounts for the projected area in any direction in the solid angle, has to be included in the integration.

We can write a similar expression for the flux $d\Phi_i$ incident over the exit pupil from a fictitious Lambertian source L_i, in the plane of the image. Then, evoking the principle of the reversibility of light, we can say that this flux, leaving the exit pupil in the direction of the image, gives rise to an image plane radiance L_i according to

$$d\Phi_i = \pi L_i \, dA_i \sin^2 \Phi'_m \tag{5.25}$$

where dA_i is the image area and θ'_m is the inclination of the marginal ray in image space. For a perfect lossless system (that is, one without reflection, absorption and scattering losses), $d\Phi_o = d\Phi_i$, so

$$L_i \sin^2 \theta'_m \, dA_i = L_o \sin^2 \theta_m \, dA_o \tag{5.26}$$

Now

$$\frac{dA_i}{dA_o} = m^2 \tag{5.27}$$

the square of the lateral magnification. We assume that the lens is aplanatic, obeying the Abbe sine condition, that is, it exhibits zero spherical abberation and coma for objects near the axis, no matter how low the F-number. (Spherical abberation is illustrated in Figure 2.7). Thus,

$$n \sin\theta_m = mn' \sin\theta'_m \tag{5.28}$$

where n and n' are the refractive indices in object and image space which we set equal to unity. Then

$$\frac{\sin \theta_{\text{m}}}{\sin \theta'_{\text{m}}} = m \tag{5.29}$$

and we get $L_i = L_o$ which shows that the radiance is conserved in a lossless imaging system. Equation (5.25) then gives for the image irradiance

$$E_i = d\Phi_i/dA_i = \pi L_o \sin^2 \theta'_{\text{m}} \tag{5.30}$$

As usual in paraxial optics, we approximate $\sin \theta'_{\text{m}}$ by $\tan \theta'_{\text{m}}$, giving

$$\sin \theta'_{\text{m}} \approx \tan \theta'_{\text{m}} = \frac{D_i}{2b} \tag{5.31}$$

where D_i is the diameter of the exit pupil and b is the image distance. By introducing the aperture number $F = f/D_i$ where f is the focal length, Equation (5.30) becomes

$$E_i = \pi L_o \left(\frac{D_i}{2b}\right)^2 = \frac{\pi L_o}{4F^2(1+m)^2} \tag{5.32}$$

With the object at infinity, $b = f$ and we get

$$E_i = \pi L_o \left(\frac{D_i}{2f}\right)^2 = \frac{\pi L_o}{4F^2} \tag{5.33}$$

Equation (5.33) or a similar form of it, is generally referred to as the 'camera equation'. It indicates that image irradiance is inversely proportional to the square of the F-(aperture) number. Therefore the diaphragm or stop openings for a lens are marked in a geometrical ratio of $2^{1/2}$.

Recall that we have assumed the object to be a Lambertian surface. For example, for a point source of radiant intensity I as the object, the flux intercepted by the entrance pupil is

$$d\Phi = I\, d\Omega = \frac{IS}{a^2} = I\pi \left(\frac{D}{2a}\right)^2 \tag{5.34}$$

where S is the area and D is the diameter of the entrance pupil, a is the object distance and where we for simplicity assume the entrance and exit pupils to have equal area S.

If we take the image area dA_i to be equal to the area of the Airy disc (see Section 4.6, Equation (4.31))

$$dA_i = \pi(\Delta r_i)^2 = 1.5\pi \left(\frac{\lambda b}{D}\right)^2 \tag{5.35}$$

we get for the image irradiance

$$E_i = \frac{d\Phi}{dA_i} = \frac{8}{3}\frac{I}{\lambda^2}\left(\frac{D}{2a}\right)^2\left(\frac{D}{2b}\right)^2 = \frac{I}{6\lambda^2}\left(\frac{m}{F^2(1+m)^2}\right)^2 \tag{5.36}$$

From this expression we see that the image irradiance is dependent on both the object and image distances. In conclusion we might say that for a Lambertian surface we can not increase the image irradiance by placing the lens closer to the object, but for a point source we can, the maximum occurring at unit magnification, i.e. when $a = b = 2f$.

5.2 INCOHERENT LIGHT SOURCES

Most light sources are incoherent, from the candle light to the Sun. They all radiate light due to spontaneous emission (see Section 5.3.1). Here we will consider some sources often used in scientific applications. These are incandescent sources, low-pressure gas discharge lamps and high-pressure gas discharge-arc lamps. They are commonly rated, not according to their radiant flux, but according to their electric power consumption.

Tungsten halogen lamps

Quartz tungsten halogen lamps (QTH) produce a bright, stable, visible and infrared output and is the most commonly applied incandescent source in radiometric and photometric studies. It emits radiation due to the thermal excitation of source atoms or molecules. The spectrum of the emitted radiation is continuous and approximates a blackbody. Spectral distribution and total radiated flux depend on the temperature, area and emissivity. For a QTH lamp, the temperature lies above 3000 K and the emissivity varies around 0.4 in the visible region.

In all tungsten filament lamps, the tungsten evaporates from the filament and is deposited on the inside of the envelope. This blackens the bulb wall and thins the tungsten filament, gradually reducing the light output. With tungsten halogen lamps, the halogen gas effectively removes the deposited tungsten, and returns it to the hot filament, leaving the inside of the envelope clean, and greatly increases lamp life. This process is called the halogen cycle. A commercial 1000 W QTH lamp have a luminous flux of up to 30 000 lm with a filament size of 5 mm × 18 mm (Oriel Corporation 1994).

Low-pressure gas-discharge lamps

In these sources an electric current passes through a gas. Gas atoms or molecules become ionized to conduct the current. At low current density and pressures, electrons bound to the gas atoms become excited to well-defined higher-energy levels. Radiation is emitted as the electron falls to a lower energy level characteristic of the particular type of gas. The spectral distribution is then a number of narrow fixed spectral lines with little background radiation. The known wavelengths determined by the energy levels are useful for calibration of spectral instruments.

High-pressure gas-discharge arc lamps

High-current-density arc discharges through high-pressure gas are the brightest conventional sources of optical radiation. Thermal conditions in the arc are such that gas atoms (or molecules) are highly excited. The result is a volume of plasma. The hot plasma emits like an incandescent source, while ionized atoms emit substantially broadened lines. The spectral distribution of the radiation is a combination of both the continuum and the line spectra. The most common sources of this type are the Xenon (Xe) and mercury (Hg) short arc lamps. Xenon lamps have colour temperatures of about 6000 K, close to that of the Sun. A commercial 1000 W Hg lamp produces a luminous flux of 45 000 lm with an effective arc size of 3 mm × 2.6 mm. A commercial 1000 W Xe lamp is even brighter with luminous flux of 30 000 lumens with an effective arc size of 1 mm × 3 mm. (Oriel Coporation 1994)

5.3 COHERENT LIGHT SOURCES

5.3.1 Stimulated emission

Figure 5.7 shows an energy-level diagram for a fictive atom or molecule (hereafter called an atom). Here only four levels are shown. Assume that the atom by some process is raised to an excited state with energy E_3. Then the atom drops to energy levels E_2, E_1 and E_0 in successive steps. E_0 may or may not be the ground state of the atom. We do not specify the type of transition from E_3 to E_2 and from E_1 to E_0, but assume that the energy difference between E_2 and E_1 is released as electromagnetic radiation of frequency v given by

$$E_2 - E_1 = hv \tag{5.37}$$

where $h = 6.6256 \times 10^{-34}$ J s is the Planck constant. This might be the situation in an ordinary light source (e.g. a discharge lamp) where the transition occurs spontaneously and the photon is therefore said to be created by spontaneous emission.

As postulated by Einstein, also another type of transition is possible: if a photon of frequency given by Equation (5.37) passes the atom it might trigger the transition from E_2 to E_1 thereby releasing a new photon of the same frequency by so-called stimulated emission. Under normal conditions known as thermodynamic equilibrium, the number of atoms in a state tends to decrease as its energy increases as shown in Figure 5.8(a). This means that there will be a larger population in the lower state of a transition than in the higher state. Therefore photons passing the atom are far more likely to be absorbed than to stimulate emission. Under these conditions, spontaneous emission dominates.

However, if the excitation of the atoms is sufficiently strong, the population of the upper level might become higher than that of the lower level. This is called population inversion and is illustrated in Figure 5.8(b). Then by passing of a photon of frequency given by

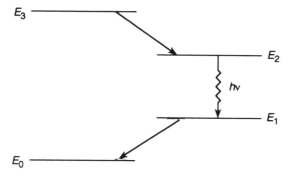

Fig. 5.7. *Energy-level diagram of a fictitious atom*

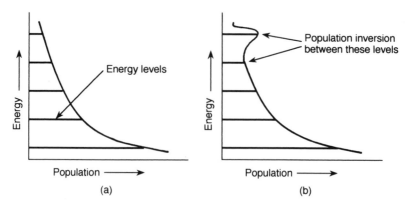

Fig. 5.8. *Distribution of populations among energy levels at (a) equilibrium, and (b) during a population inversion*

Equation (5.37), it will be more likely to stimulate emission from the excited state than to be absorbed by the lower state. This is the condition that must be obtained in a laser and the result is laser gain or amplification, a net increase in the number of photons with the transition energy. Light amplification by the stimulated emission of radiation therefore has given rise to the acronym 'laser'. Laser gain is proportional to the difference between the chance of stimulated emission and the chance of absorption. Therefore the population of both the upper and lower levels of the laser transition are important. Thus if laser action is to be sustained, the lower level must be depopulated as the upper level is populated or the population inversion will end. That is indeed what happens in some pulsed lasers. Stimulated emission has the same wavelength as the original photon and it is in phase (or coherent) with the original light.

In the description given above, four energy levels are involved. This is the best condition for laser action and is called a four-level laser. But also three-level lasers exists which is the case when e.g. the lower transition level is the ground state. To maintain population inversion, it is easily realized that the lifetime of the E_2-level should be as long as possible and the lifetime of the E_1-level should be as short as possible. The process of raising the atom to the E_3 excited level is called pumping. The pumping mechanism is different for different laser types.

In the description given above, we have assumed laser transition between energy levels E_2 and E_1 only. Usually, stimulated emission can be obtained between many different energy levels in the same laser medium. Dependent on the construction of the laser, one can obtain lasing from a single transition or from a multitude of transitions.

Laser amplification can occur over a range of wavelengths because no transition is infinitely narrow. The range of wavelengths at which absorption and emission can occur is broadened by molecular motion (Doppler broadening) vibrational and rotational energy levels, and other factors.

To be more specific, let us consider the most familiar of all lasers, the helium–neon (He–Ne) laser.

Figure 5.9 shows the construction of a typical He–Ne laser. Inside a discharge tube is a gas mixture of helium and neon. Typically the mixture contains 5 to 12 times more helium than neon. By applying voltage to the electrodes, the resulting electric field will

Fig. 5.9. *He-Ne laser. (Hecht & Zajac, Optics, © 1974, Addison-Wesley Publishing Company, Inc., Reading, Massachusetts. Figure 14.31. Reprinted with permission)*

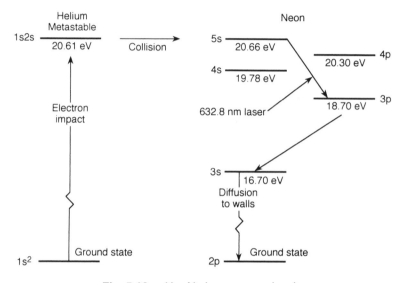

Fig. 5.10. *He–Ne laser energy levels*

accelerate free electrons and ions inside the tube. These collide with helium atoms raising them to a higher energy level. By collision between helium and neon atoms, the latter are raised to a higher energy level. This constitute the pumping process. The neon atoms, which constitute the active medium, return to a lower energy level and the energy difference is released as electromagnetic radiation.

Figure 5.10 shows an energy-level diagram for an He–Ne laser emitting red light. Excited helium atoms in the 1s2s state transfer energy to neon atoms in the ground state by collisions, exciting the neon atoms to the 5s excited state. By returning to the 3p state the energy difference is released as light of wavelength 632.8 nm.

Population inversion alone is sufficient to produce 'light amplification by the stimulated emission of radiation', but the result is only a coherent monochromatic light bulb. In fact, population inversion is observed in the atmosphere of Mars. To get light oscillation however, the discharge tube is enclosed in an optical cavity or resonator which is two mirrors facing each other as in Figure 5.9. The result is that the light is reflected back and forth through the tube, stimulating emission again and again from neon atoms. Emission in other directions is lost out of the laser medium and the light is concentrated in a beam oscillating back and forth between the mirrors. The optical cavity therefore acts as an oscillator or feedback amplifier and is an essential part of a laser.

Below we give a short description of other lasers. Numerous lasers and laser media have been demonstrated in the laboratory. Here we concentrate on lasers which are available commercially. For further details, the excellent book by Hecht (1992) is highly recommended.

There are many potential criteria for classifying lasers, but the two most useful ones are the type of active medium and the way in which it is excited (pumped). Usual practice is to group most devices as gas, liquid, solid-state or semiconductor lasers. A few important lasers are exceptions. Liquid- and solid-state lasers are pumped optically, i.e. by means of a flashlamp or another laser. Semiconductor lasers are excited when charge carriers in a semiconductor recombine at the junction of regions doped with n- and p-type donor materials. Gas lasers can be pumped in various ways, including discharge excitation (cf. the He–Ne laser), radio frequency (RF) excitation, chemical and optical excitation and also by gas expansion (gas dynamics).

5.3.2 Gas lasers

Helium–neon

Among the first lasers demonstrated and the first gas laser (Javan *et al.* 1961). The 632.8 nm line is the most important because it can give up to about 50 mW continuous wave (c.w.). Green, yellow, orange and multiline versions are being offered commercially. Advantages: Output beam of low divergence and high coherence.

Noble gas ion lasers

Emit on ionized rare gas lines. Pumping: Electrical discharge. The most important is argon with strong lines in the blue-green and weaker lines in the ultraviolet and near-infrared. The 514.5 nm line is the strongest in larger water-cooled lasers while the 488.0 nm line is the strongest in small air-cooled models. Another type is Krypton lasers.

Advantages: Their ability to produce c.w. output of a few milliwatts to tens of watts in the visible and up to 10 W in the ultraviolet.

Helium–cadmium lasers

Emit on lines of ionized metal vapours. Electric discharge pumping. C.w. output up to about 150 mW at 442 nm or powers to about 50 mW at 325 nm.

Advantages: Offer short wavelength at moderate output power which can be focused to a small spot.

Carbon dioxide lasers

Pumping: Electric discharge, RF or gas-dynamic. Transitions between vibrational levels. Infrared radiation between 9 and 11 µm. Several distinct types. Can produce c.w. output powers from under 1 W for scientific applications to many kilowatts for materials working. Can generate pulses from the nanosecond to millisecond regimes. Custom-made

CO_2 lasers have produced c.w. beams of hundreds of kilowatts for military weapon research or nanosecond pulses of 40 kJ for research in laser-induced nuclear fusion. Advantages: No other commercial laser can generate as intense c.w. output.

Chemical lasers

Pumping by means of chemical reaction. Three most important media: hydrogen fluoride, deuterium fluoride and iodine. Emits at wavelengths between 1.3 μm and 4.2 μm. Military research has demonstrated building-sized lasers that have generated nominally c.w. outputs to a couple of megawatts. Commercial devices produce much lower powers.

Copper and gold vapour lasers

Emit in or near the visible region on lines of neutral metal vapor. Pumping: Electric discharge. Operate as pulsed lasers only. Commercial copper vapour lasers can emit over 100 W in the green and yellow, gold vapour lasers can generate several watts in the red. Advantages: High power and high efficiency in the actual wavelength region with repetition rates of several kilohertz.

Eximer lasers

Eximer is a contraction of 'excited dimer', a description of a molecule consisting of two identical atoms which exists only in an excited state, e.g. He_2 and Xe_2. Since the ground state does not exist, population inversion is obtained as long as there are molecules in the excited state. Pumping: Electric discharge transverse to the gas flow. Most important media: rare gas halides such as: argon fluoride, krypton fluoride, xenon fluoride and xenon chloride. Emit powerful pulses (average power of up to 100 W) lasting nanoseconds or tens of nanoseconds in or near the ultraviolet.

Advantages: Very high gain. No other commercial laser can generate such high average power at such a short wavelength.

Nitrogen lasers

Pumping: Electric discharge. Can produce nanosecond or subnanosecond pulses (average power of a few hundred milliwatts) of wavelength 337 nm.

Advantages: Low-cost. So simple to build that it were once featured in the 'Amateur Scientist' column of *Scientific American*.

5.3.3   Liquid lasers

Dye lasers

The discussion of liquid lasers almost invariably starts and ends with the dye laser. The active medium is a fluorescent organic dye dissolved in a liquid solvent.

Pumping: Optical, with a flashlamp or (more often) with an external laser. The output wavelength can be tuned from the near-ultraviolet into the near-infrared. Dye lasers can be adjusted to operate over an extremely narrow spectral bandwidth and can also produce ultrashort pulses, much shorter than a picosecond.

Disadvantages: Very complex. Tuning wavelength across the visible spectrum requires several changes of dye. Complex optics are needed to produce either ultra-narrow linewidth or picosecond pulses.

5.3.4   Semiconductor diode lasers

Figure 5.11(a) shows a semiconductor diode consisting of a p-type material, i.e. a material doped with impurities to produce holes (or equivalently, accept electrons) and an n-type material, i.e. a material doped with donors of excess electrons. Between the two materials is a junction, or active layer. Application of a voltage to the terminals, as shown in the figure, creates a local excess of minority carriers on each side of the junction. At the junction, excess electrons fall to lower energy states, i.e. holes in the valence band. This process is called electron–hole recombination. By passing a high drive current through the diode, population inversion is produced at the junction, generating photons of energy roughly equal to the bandgap, the energy difference between the conduction and valence bands. Lower drive currents give incoherent spontaneous emission and the diode acts as a light emitting diode (LED). The optical feedback comes from reflective structures which form the diode laser resonator, see Figure 5.11(b).

Much of the work on semiconductor diode lasers has been on the design of junction structures. It suffices here to say that the so-called double-heterojunction (or double-heterostructure) laser with the active layer formed as a waveguide (see Section 12.4) by means of variable refractive index has been the most successful device.

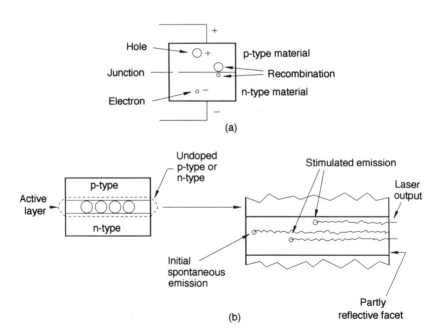

Fig. 5.11.   *Simplified operating principle of a diode laser: (a) forward bias produces current flow and recombination; (b) stimulated emission in active layer*

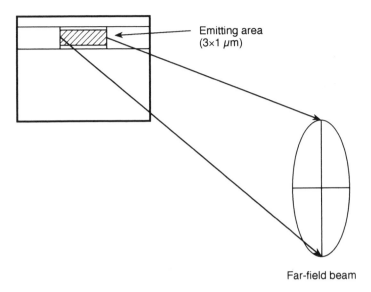

Emitting area
(3×1 µm)

Far-field beam

Fig. 5.12. *Elliptical divergence of light from a narrow-stripe diode laser*

The bandgap depends on the crystalline structure and chemical composition of the semiconductor material. Binary compounds such as GaAs or InP have a fixed bandgap and wavelength. Adding a third element (such as aluminium) changes the bandgap. Such mixtures are often labelled in the manner $Al_xGa_{1-x}As$, where the x denotes the fraction of one element, in this case aluminium.

Individual diode lasers are tiny blocks of semiconductor material less than a millimetre on each side. Most emit light from the edge of the wafer, but some newer types emit light from the surface. Also arrays of many parallel laser stripes are produced. In practice, up to about 200 laser stripes can be fabricated on a 1 cm wide bar, with total output power of 10 W c.w.

With a very short cavity and a very small emitting area, a diode laser produces a rapidly diverging beam as shown in Figure 5.12. Divergence is much larger perpendicular to, than in the plane of, the active layer because of diffraction. However, the beam can be made circular by means of cylindrical optics.

Semiconductor diode lasers can be divided into short-wavelength (below about 1.1 µm) and long wavelength lasers. The lasers belonging to the first group are listed in Table 5.2.The huge volume of low power 780 nm GaAlAs lasers are produced for compact-disc (CD) players.

Semiconductor diode lasers with wavelength longer than 1.1 µm are used primarily for fibre-optic communications. The material system used for commercial lasers is InGaAsP which can be made for a broad range of wavelengths. Work has been concentrated on 1.31 µm where silica step-index single mode optical fibres have zero wavelength dispersion and loss about 0.5 dB per kilometre, and 1.55 µm where silica fibres have their lowest loss, about 0.15 dB/km, see Section 12.4.2. Most fibre-optic systems today work on 1.31 µm. Advantages: Very compact, low power consumption (do not require high voltages to drive a gas discharge), low price.

Table 5.2   *Diode laser wavelengths under 1 μm and power levels available commercially. (From Hecht, J. (1992), The Laser Guide Book (2nd edn), McGraw-Hill, N.Y.). (Reproduced by permission of McGraw-Hill, Inc)*

Nominal wavelength (nm)	Compound	Maximum continuous-wave power (single element)
635	InGaAlP	3 mW
660	InGaAlP	3 mW
670	$Ga_{0.5}In_{0.5}P$	10 mW
750	GaAlAs	8 mW
780	GaAlAs	35 mW
810	GaAlAs	100 mW single
		10 W linear array
		60 W quasi-continuous-wave (pulsed) array
		1500 W quasi-continuous-wave stacked array
830	GaAlAs	150 mW
850	GaAlAs	100 mW
880 or 895	GaAlAs	Pulsed only
905	GaAs (nominal)	Pulsed only
910	InGaAs	Pulsed only
980	InGaAs	50 mW

5.3.5   Solid-state lasers

A solid-state laser is one in which the active medium is a non-conductive solid, a crystalline material, or glass doped with a species that can emit laser light. In crystalline or glass solid-state lasers, the active species is an ion embedded in a matrix of another material, generally called the 'host'. It is excited by light from an external source.

Neodymium lasers

The active medium is triply ionized neodymium in a crystal or glass matrix. The most common host materials are: yttrium lithium fluoride (YLE), phosphate glass, gadolinium scandium gallium garnet (GSGG), silicate glass and yttrium aluminum garnet (YAG) with wavelengths ranging from 1047 to 1064 nm (Nd-YAG). Neodymium lasers can generate c.w. beams of a few milliwatts to over a kilowatt, short pulses with peak powers in the gigawatt range, or pulsed beams with average powers in the kilowatt range.

Advantages: A very versatile laser that can be doubled, tripled and quadrupled by means of harmonic generation and generate short pulses with high power by means of Q switching and modelocking, see Section 5.3.7.

Ruby lasers

The first laser demonstrated (Maiman 1960). Ruby laser rods are grown from sapphire (Al_2O_3) doped with about 0.01 to 0.5 percent chromium. Emits at 694.3 nm. C.w operation has been demonstrated in the laboratory, but is difficult to achieve. Oscillators can produce millisecond pulses of 50–100 J. Require active cooling.

Tunable vibronic solid-state lasers

Tunable wavelength due to operation on 'vibronic' transitions in which the active medium changes both electronic and vibrational states. Commercial lasers made from: alexandrite (chromium-doped $BeAl_2O_4$) which can be tuned between 701 and 826 nm. Titanium-doped sapphire (Al_2O_3), tunable from 660 to 1180 nm, and cobalt doped magnesium fluoride between 1750 and 2500 nm (wavelength ranges given at room temperature). Can be operated both c.w. and pulsed. Output power depends on wavelength. Commercial pulsed alexandrite lasers can generate average powers to 20 W, Ti sapphire reaches several watts c.w.

Advantages: Ti sapphire has the broadest tuning range of any single conventional laser medium. (Dye lasers can be tuned across a broader range only by switching dyes.)

Fiber lasers and amplifiers

The fibre laser is a miniaturization of solid-state lasers. Interest has concentrated on fibre amplifiers to replace conventional electro-optic repeaters used in fibre-optic systems. Such repeaters detect a weak optical signal, convert it into electronic form, amplify and process the electronic signal, and use it to drive a laser transmitter.

The basic concept is shown in Figure 5.13. A fibre is made from a solid-state material (typically a glass) doped with an ion which emits at the desired wavelength λ_1. It is illuminated from one end by a weak signal at λ_1 and a stronger steady beam at a second wavelength λ_2 which excites the ion in the fibre to the upper laser level. As the weak signal passes through the fibre, it stimulates emission from excited ions at λ_1. Interest has centred on $\lambda = 1.3\,\mu m$ with neodymium as the laser ion and $\lambda = 1.54\,\mu m$ with erbium. For practical applications, diode lasers are used for pumping.

Other solid-state lasers

A lot of solid-state laser materials have been demonstrated in the laboratory. Here we mention the erbium–glass laser and the crystalline erbium laser (Er-YAG). The first emits at $1.54\,\mu m$ and is therefore a candidate for eye-safe laser range finders (The $1.06\,\mu m$ wavelength of Nd-YAG poses a serious eye hazard.) The most important line of Er-YAG

Fig. 5.13. *Operating principle of a fibre amplifier*

is at 2.94 μm which is absorbed strongly by water, so it leaves a thinner damaged layer between healthy tissue and the zone removed by surgery. The absorption is so strong that it can be used to cut bone.

5.3.6   Other lasers

Here we mention two types:

The free electron laser

The central idea is to extract light energy from electrons passing through a magnetic field with periodic variations in intensity and directions. It is therefore not based on stimulated emission but promises extremely high powers and exceptionally broad tunability, from microwaves to X-rays.

X-ray lasers

Visible and near-ultraviolet lasers operate on electronic transitions in the outer or valence shells of atoms. Transitions from outer to inner shells involve much more energy, thus producing X-rays. However, conditions for producing population inversion on such inner-shell transitions are extremely difficult to obtain. Two methods have been demonstrated by the Lawrence Livermore National Laboratory. One used a nuclear bomb explosion, the other used short, intense pulses from high-energy lasers built for fusion research.

5.3.7   Enhancements of laser operation

A description of lasers is not complete without mentioning some techniques that can enhance their operation. Here we give a short introduction to methods for wavelength enhancements, i.e. laser line narrowing and alteration of the laser wavelength, and changing of the pulse length.

Line narrowing

As indicated in Figure 5.14, the gain profile of even nominally monochromatic lasers normally spans several longitudinal modes, each with slightly different wavelength. The basic idea of line narrowing is to insert into the laser cavity optical elements, which restricts oscillations to a range of wavelengths so narrow that it includes only a single longitudinal mode. The commonest line-narrowing component is the Fabry–Perot etalon, typically a transparent plate with two reflective surfaces forming a short resonator that can be inserted within the laser cavity. In this way, frequency bandwidths as low as 500 kHz are obtainable from commercial lasers.

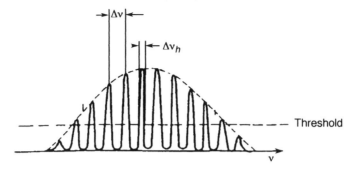

Fig. 5.14. *Spectral distribution of a He-Ne laser Δv = resonator mode spacing, Δv_h = resonator linewidth*

Wavelength alteration

Techniques for changing the wavelength from a laser include harmonic generation, parametric oscillation and Raman shifting. The method in most practical use is harmonic generation. This is based on the nonlinear interactions between light and matter (usually a non-linear crystal) which can generate harmonics at multiples of the light-wave frequency. The magnitude of the non-linear effect is proportional to the square of the incident power. Therefore, for most practical applications only the second, third and sometimes the fourth harmonics are produced. Conventionally, the laser beam makes a single pass through the crystal (usually potassium dihydrogen phosphate, KDP) which is placed outside the laser cavity. The commonest use of harmonic generation is with the 1064 nm Nd-YAG laser producing the 532 nm second, the 355 nm third and the 266 nm fourth harmonic. Dye laser output is often frequency doubled to obtain tunable ultraviolet light and also GaAlAs semiconductor lasers to give blue light.

Three techniques which operate by interacting with light inside the laser cavity for producing short pulses are in widespread use. These are Q-switching, cavity dumping and modelocking.

Q-switching

Like any oscillator, a laser cavity has a quality factor Q, defined as

$$Q = \frac{\text{energy stored per pass}}{\text{energy dissipated per pass}}$$

Normally, the Q-factor of a laser cavity is constant, but if the Q-factor is kept artificially low, energy will gradually accumulate in the laser medium because the Q-factor is too low for laser oscillation to occur and dissipate the energy. If the Q-factor is increased abruptly, the result is a large population inversion in a high-Q cavity, producing a high-power burst of light, a few nanoseconds to several hundred nanoseconds long, in which the energy is emitted. This rapid change is called Q-switching. Figure 5.15 shows the three basic variations on Q-switching. The first uses a rotating mirror or prism as the rear cavity mirror. Periodically the rotating mirror passes through the point where it is properly

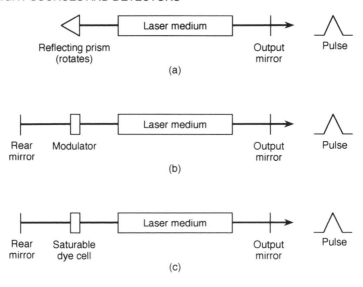

Fig. 5.15.   *Types of Q switching: (a) rotating mirror or prism; (b) active modulator; (c) passive*

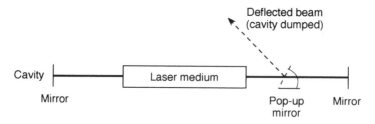

Fig. 5.16.   *The basic concept of cavity dumping*

aligned with the output mirror, causing cavity Q to increase abruptly and producing an intense Q-switched pulse. The second is insertion of a modulator (usually electro-optic or acousto-optic devices) into the cavity, blocking off one of the cavity mirrors. The third variation is insertion of a non-linear lossy element into the cavity that becomes transparent once intra-cavity power exceeds a certain level.

Cavity dumping

The basic idea of cavity dumping is to couple laser energy directly out of the cavity without having it pass through an output mirror. In this case, both cavity mirrors are totally reflective and sustain a high circulating power within the laser cavity. The concept is illustrated in Figure 5.16 where a mirror pops up into the cavity and deflects a pulse with length close to the cavity round-trip time. In practice, cavity dumping is done with other arrangements. Cavity dumping can be used with c.w. lasers which do not store energy in the upper excited levels and hence cannot be Q-switched. Cavity dumping of a c.w. laser can generate pulses of 10–50 ns at repetition rates of 0.5–5 MHz.

Modelocking

Modelocked pulses can be visualized as a group of photons clumped together and aligned in phase as they oscillate through the laser cavity. Each time they hit the partially transparent output mirror, part of the light escapes as an ultra-short pulse. The clump of photons then makes another round trip before the next pulse is emitted. Thus the pulses are separated by the cavity round-trip time $2L/c$, where L is the cavity length and c the speed of light. The physics is much more complex and well beyond the scope of this book. Modelocking can generate pulses in the picosecond regime. It requires a laser that oscillates in many longitudinal modes and will therefore not work for many gas lasers with narrow emission lines. However, it can be used with argon or krypton ion lasers, semiconductor lasers and dye lasers. The pulse length is inversely proportional to the laser's oscillating bandwidth, so dye lasers can generate very short pulses because of their broad gain bandwidths.

Q-switching, cavity dumping and modelocking may be used in combination.

5.3.8 Applications

Since its invention in 1960, the laser has found numerous applications. With regard to optical metrology, the He–Ne laser and also the argon ion laser have long been a standard choice for holography and interferometry. In recent years also single narrow-stripe semiconductor diode lasers (GaAlAs) operating in a single longitudinal mode have been applied in holography. For pulsed holography, the ruby laser and the neodymium laser are most common.

Most of the laser types described above have been used in numerous different types of applications. Below we mention some of the more successful application areas.

- *Compact-disc players.* Designed around the 780 nm beam of a few milliwatts from GaAlAs semiconductor diode laser. Nearly 20 million CD audio players are sold each year. The most successful commercial single applications of lasers.

- *Writing and recording.* In laser printers, the beam scans a photoconductive drum, discharging the electrostatic charge held by the surface at points where the beam is 'on'. The resulting pattern is then printed on paper by a copierlike process. Lasers also can encode data as a series of dots on light-sensitive discs for computer data storage. He–Ne lasers initially were used for these applications, but today all but the most expensive high-speed systems use semiconductor lasers. For high-speed systems, He–Cd are the most common lasers.

- *Reading and scanning.* The biggest single application for He–Ne lasers has been in bar-code readers for supermarket checkout. The beam repetively scans a well-defined pattern. This application requires red light and good beam quality, so He–Ne lasers are likely to remain dominant in the near future.

- *Medicine.* Surgery is performed with the 10 μm line from CO_2 lasers with output power in the 50 W range. Neodymium lasers are standard tools for cataract surgery (The membrane inside the eye becomes cloudy). Unlike the CO_2, light from Nd lasers can be carried through optical fibres allowing use in endoscopes for e.g. gall-bladder surgery. Er-YAG is a promising laser for surgery.

- *Materials working.* The CO_2 laser is used for materials working, primarily cutting and welding of metals and non-metals and heat treating of metals. It is complemented by Nd-YAG which is better for drilling, spot-welding and laser marking.

- *Range finding.* The largest single use of Nd-YAG lasers probably is as military range finders and target designators. Their pulses pose a serious eye hazard to friendly troops and Erbium glass is an alternative.

- *Spectroscopy and analytical chemistry.* The ability to tune the dye laser wavelength and light emission to a narrow spectral bandwidth and the generation of ultra-short picosecond and femtosecond pulses makes dye lasers extremely useful in this reseach area. The tuneable Ti-sapphire laser is an alternative in the near-infrared regime.

- *Communications.* The InGaAsP semiconductor diode laser is used for fibre-optic communication systems, see Section 12.4.3. The 511 nm line of the copper vapour laser is suitable for underwater transmission.

5.3.9 The coherence length of a laser

Although the laser light has a well-defined wavelength (or frequency), it has nevertheless a certain frequency spread. By spectral analysis of the light, it turns out that it consists of one or more distinct frequencies called resonator modes, separated by a frequency difference equal to

$$\Delta v = c/2L \tag{5.38}$$

where $c =$ the speed of light and L is the distance between the laser mirrors, i.e. the resonator length. Thus, the spectral distribution of the light from a multimode He–Ne laser is typically as given in Figure 5.14.

Now, assume that we apply a laser with two resonator modes as the light source in the Michelson interferometer in Figure 3.14. We then have two wave fields u_1 and u_2 with frequencies

$$v_1 = \frac{c}{\lambda_1} \tag{5.39a}$$

$$v_2 = \frac{c}{\lambda_2} = v_1 + \Delta v \tag{5.39b}$$

u_1 will interfere with itself but not with u_2, and vice versa, and the total intensity thus becomes (see Equation (3.33)):

$$I(l) = 2I\left(1 + \cos\frac{2\pi v_1 l}{c}\right) + 2I\left(1 + \cos\frac{2\pi v_2 l}{c}\right) \tag{5.40}$$

where l is the path length difference and where we have assumed u_1 and u_2 to have equal intensity I. Equation (5.40) can be rearranged to give

$$I(l) = 4I\left[1 + \cos\frac{2\pi(v_1 - v_2)l}{2c}\cos\frac{2\pi(v_1 + v_2)l}{2c}\right] \tag{5.41}$$

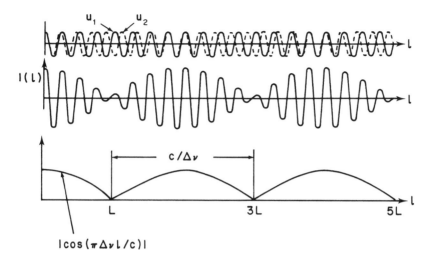

Fig. 5.17. *(a) Field amplitudes u_1 and u_2, (b) intensity distribution, and (c) temporal degree of coherence (visibility function) resulting from two resonator modes of a laser*

We see that the interference term is the same as that obtained with a light source with the mean frequency $(v_1 + v_2)/2$, but multiplied (modulated) by the factor

$$\cos\frac{2\pi\Delta vl}{2c}$$

This means that each time

$$\frac{\Delta vl}{2c} = \frac{n}{2} \quad \text{for} \quad n = 0, 1, 2, \dots$$

i.e.

$$l = n\frac{c}{\Delta v} = n2L \tag{5.42}$$

the contrast or visibility of the interference pattern will have a maximum. The visibility and therefore the temporal degree of coherence (see Section 3.2) in this case is therefore equal to

$$|\gamma(\tau)| = \left|\cos\frac{\pi\Delta vl}{c}\right| = \left|\cos\frac{\pi l}{2L}\right| \tag{5.43}$$

The path-length difference corresponding to the first minimum in the visibility function Equation (5.43), is called the coherence length (see Section 3.2). This is illustrated in Figure 5.17. If more than two resonator mode had been taken into account, the result would have been essentially the same, i.e. the same locations of the minima, but with a more steeply varying visibility function $|\gamma(\tau)|$. From this we conclude that when applying a laser in interferometry, the path length difference should be nearly zero or an integer number of twice the resonator length.

Table 5.3 *Nominal characteristics of holographic emulsions. From Vest 1979. Reproduced by permission of Wiley*

Type	Thickness (μm)	Resolution (lines/mm)	Nominal[a] sensitivity (μJ/cm^2)	Wavelength (nm)
Kodak emulsions				
649F	17	2000	70	632.8
120-02	6	2000	30	632.8
125	3	1250	2	441.6
			5	514.5
131	9	1250	0.5–0.8	632.8
Agfa-Gevaert emulsions				
8E75	7	5000	10	632.8
10E75	7	2800	2	632.8
			1.8	514.5
8E56	7	5000	15	476
			25	521
10E56	7	2800	1.4	476
			1.9	521
14C70		1500	0.3	700
14C75		1500	0.3	700
Ilford emulsion			5	632.8
He–Ne 1	—			

[a]These values may be used for estimation of exposure times. There is considerable variation in reported data, and the sensitivites may correspond to various densities in the range 0.4 to 1.0. Also, there are variations in reported values for resolution.

5.4 HOLOGRAM RECORDING MEDIA

5.4.1 Silver halide emulsions

Apart from the laser, the photosensitive material for recording the hologram is the most important part of a holographic set-up. When making a hologram recording such as that illustrated in Figure 6.1, the photosensitive material must resolve the complicated intensity distribution resulting from interference between the object and reference waves. The mean frequency in this pattern is given by the mean angle between these two waves. Owing to space limitations, this angle in practice is about 20° and greater. From Figure 3.5 we can see that this angle corresponds to a spatial frequency of about 500 lines/mm. Normally a resolution of 1000 to 2000 lines/mm is desirable. This criterion is met by several silver halide emulsions (Vest 1979). A number of them are listed in Table 5.3.

In the description of hologram recording in Section 6.1 we assume a linear relation between the amplitude transmittance t and the exposure E of a hologram, where E is the intensity times the exposure time, i.e. the energy density per unit area. This assumption is not strictly true. A typical $t - E$ curve for a holographic emulsion is shown in Figure 5.18. Another cmmon transmission characteristic of films is the Hurter–Driffield curve, which

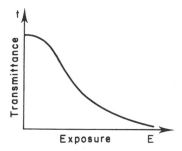

Fg. 5.18. *t − E curve for holographic emulsion*

is a density versus $\log E$ curve. The density D is defined as

$$D = \log\frac{1}{|t|^2}$$

where the square of the amplitude transmittance is called the intensity transmittance. Density is a common parameter for ordinary photography since the eye detects brightness differences on an approximately logarithmic scale. Photographic films are often characterized by the slope γ of the linear portion of the D–$\log E$ curve.

In Section 6.3 we shall see that to obtain maximum diffraction efficiency of an amplitude hologram, the bias transmittance t_b should be equal to 0.5 which corresponds to $D = 0.6$. Because of the non-linearity of the $t − E$ curve it is advantageous to have t_b slightly lower (i.e. D slightly higher) than this value. In ordinary photography the bias density is centred at the linear portion of the D–$\log E$ curve which does not coincide with the linear portion of the $t − E$ curve. A properly recorded amplitude hologram therefore looks underexposed compared to a photograph.

Holograms can be recorded in many other materials such as dichromated gelatin, photoresist, photopolymers, photochromic materials, ferroelectric crystals and thermoplastic films. In Section 5.4.2 we describe in some detail the thermoplastic film.

5.4.2 Thermoplastic film

A thermoplastic, e.g. Staybelite Ester 10, is not photosensitive and must therefore be combined with a photoconductor in a film structure (Urbach and Meier 1966). The system is usually built up from a substrate of glass upon which is coated a conducting layer of, for example, tin oxide. On this is deposited a photoconductor such as polyvinylcarbazole sensitized with trinitro-9-fluorenone and on top of this is deposited a thermoplastic layer. The recording technique consists of a number of steps, beginning with establishing a uniform electrostatic charge on the surface of the thermoplastic with a corona discharge assembly. This charge is capacitively divided between the photoconductor and the thermoplastic layers and upon subsequent exposure the interference pattern causes the photoconductor to discharge in a spatial pattern dependent on the exposure. However, this does not cause any variation in the charge on the thermoplastic. This is accomplished by recharging the surface uniformly, which increases the charge in the illuminated areas. The thermoplastic is then heated to the softening temperature, allowing electrostatic forces to deform it so that it becomes thinner at illuminated areas and thicker elsewhere. Cooling quickly to room temperature, the deformations are frozen in, resulting in

Centring magnets

Glass window

Target

g_4

g_3

g_2

g_1

Target connection ring

Cathode

Horizontal and vertical deflection coils

Focusing coil

(a)

Transparent film of tin oxide

a   b

Glass window

Light-sensitive semiconductor material

Connection ring

To preamplifier

R

+40V

(b)

Fig. 5.19. *(a) The vidicon tube. (b) Construction of the target. (c) A target element*

a hologram with thickness variations, i.e. a phase hologram. Reheating the thermoplastic to a higher temperature tends to restore it to its original state. Thus the material has a write–erase recycling capability. This is quite a complicated procedure, but now complete camera units are becoming commercially available, giving a hologram ready for reconstruction within 5 s of the exposure.

A peculiar feature of the thermoplastic film is that it has a band-limited spatial frequency response centred at about 1000–2000 lines/mm.

Target element P

Resistance of the
semiconductor element

Electron beam

Tin oxide

Capacitance of the·
semiconductor element

To preamplifier

R

+40V

Fig 5.19. (c)

5.5 THE VIDICON TUBE

Figure 5.19(a) shows the construction of a standard TV vidicon tube (Von Wezel 1981; Biberman and Nudelman 1971). It consists of an electron gun which produces an electron beam that scans the target. g_1 is the Wehnelt cylinder by means of which the beam current can be set, g_2 is the anode, g_3 is the focusing electrode and g_4 is a piece of fine-mesh metal gauze which makes the electric field between the target and g_3 as homogeneous as possible. Figure 5.19(b) shows the target construction. The glass window is provided with two films: 'a' is a transparent film of tin oxide connected to the connection ring around the vidicon, 'b' is a film of photoconductive semiconductor material (antimony trisulphide in the case of a standard vidicon).

The conductive tin oxide is supplied with, for example, $+40$ V through a resistor R. The potentials on the different electrodes are adjusted so that an electron is decelerated more than accelerated so it will never actually reach the target. Since the electron beam forms a direct connection with the cathode, the back of film 'b' is brought to cathode potential each time the electron beam passes.

Let us now consider a small element P of the target (Figure 5.19(c)). This may be regarded as a small capacitor which is slowly discharged through the semiconductor material after the electron beam has passed and which is, through R, recharged to 40 V the moment the electron beam passes. This charging current is called dark current I_d when the vidicon is receiving no light. At 40 V, I_d is typically 20 nA. When light falls on P, the resistance of the semiconductor material decreases and the capacitor is discharged more quickly. This means that a larger charging current is needed and this forms the signal current I_s. To see that this will result in a signal current proportional to the illumination of each target element, it must be appreciated that the target material has a very high longitudinal resistance to prevent leakage of the charge from the target.

Camera tubes are generally rated in terms of signal current for a given light level or radiation input level. This quotient, signal current divided by incident flux (measured in lumens) is called the sensitivity (or responsivity) of the tube. The slope of the curve giving I_s as a function of illumination (measured in lux) in a log–log scale is referred to as the γ of the tube, just in the same manner as for a photographic film. For a standard vidicon, γ can be varied by varying the target voltage. The signal-to-noise ratio of a camera tube is ideally

Table 5.4   *A comparison of some camera tubs. From van Wezel 1981*

	Vidicon	Plumbicon	Silicon target vidicon	Chalnicon Pasecon	Saticon	Newvicon
Designed by:	RCA	Philips	Bell	Toshiba	Hitachi	Matsushita
in:	1951	1963	1960	1972*77	1973	1974
Photosensitive target	Sb_2S_3	PbO	Si	CdSe	SeAsTe	ZnSe/ZnCdTe
Picture-size (1 in)	——————————		9.6 × 12.8 mm		——————	
Sensitivity (μA/l m) (visible light)	variable	400	900	1500	350	1200
γ	0.7	1	1	1	1	1
Signal current (μA)	0.2	0.2	0.2	0.2	0.2	0.2
Dark current (nA)	20	<1	10	<1	<1	6
Lag (%) residual signal current after 60 ms)	20	2	7	10	3	10
Modulation depth at 5 MHz (%)	60	50	40	60	60	55
Signal-noise ratio (linear, dB)	45	47	45	>45	>45	>45
Disturbing co-effects	lag: dark current	——— blooming at highlights ———	dots; dark current	lag	max. temp 50·C	dark current
Target burns	moderate	good	very good			good
Infrared sensitivity	moderate	moderate	very good	sufficient	moderate	good
Type						
EEV 1 inch	7262A	P 8021	P 8120			
2/3 inch		P 8160				
RCA 1 inch	8507 A		4532 H		BC 4395	4906
2/3 inch	8844		4833		BC 4390	4904
Philips 1 inch	XQ 1032	XQ 1070	XQ 1402			XQ 1440
2/3 inch	XQ 1270	XQ 1427				XQ 1274
Heimann 1 inch	XQ 1291	XQ 1352	XQ 1205	XQ 1461		
2/3 inch	XQ 1311		XQ 1313	XQ 1451		

Deflection yokes for   1 inch tubes: Phillips AT 1102
2/3 inch tubes: Philips K V 198B
K V 12

determined only by the signal-to-noise ratio associated with the photoconductive current resulting from the image on the target. In general, the limitations for vidicons result from amplifier noise. Thus, vidicons having the highest sensitivity also have the higher signal-to-noise ratio at a given light level.

Another characteristic of a camera tube is the lag effect which is the residual signal that is observed after the light is removed from the target. This is because a single pass of the electron beam might not be sufficient for a full recharge of the semiconductor capacitance. Thus, lag increases with increasing capitance of the target, but is also influenced by other effects.

Since the appearance of the vidicon on the market in 1951, many manufacturers have tried to improve upon it. As almost all weak image points can be traced back to the target photoconductive layer, the newcomers differ from the vidicon in the use of layer materials only. Table 5.4 makes a comparison of the best-known types.

The electron beam scans the 9.6×12.8 mm target in 625 horizontal lines every 1/25 second. This gives a vertical resolution of $625/2 \div 9.6 \sim 32$ lines/mm. The duration of line becomes $1/25 \div 625 = 64\,\mu s$. To have equal resolution in the horizontal direction, $625/2 \times 12.8/9.6$ vertical lines should be resolved during this time. This gives a frequency equal to

$$\frac{625 \times 12.8}{2 \times 9.6 \times 64}\,\text{MHz} = 6.5\,\text{MHz}$$

The nominal frequency bandwidth in TV transmission is 5.5 MHz. Because some of the lines are lost in synchronization and flyback of the electron beam, the above figures are not strictly correct, see Section 5.8. A resonable value of the resolution of a standard vidicon camera can therefore be taken to be about 30 lines/mm both in the horizontal and vertical directions.

5.6 THE CCD CAMERA

5.6.1 Operating principles

Until the mid 1960s, electronic devices for the pick-up of optical images were in the form of vacuum-type camera tubes. During the 1960s solid-state arrays with individual photoconductor elements connected to X–Y conductors and sequentially activated by voltages from thin-film shift registers were developed. The resulting images were however severely limited by response non-uniformities and other form of spatial noise associated with the X–Y readout techniques.

Workers at the Bell Telephone Laboratories (Boyle and Smith 1970) presented a new semiconductor device concept based on the manipulation of charge packets rather than the modulation of electric currents. Below we give a brief description of this concept. For a more thorough description, the article by Barbe and Campana (1977) is recommended.

A CCD is essentially a series of metal oxide semiconductor (MOS) capacitors. Figure 5.20 shows a simplified sketch of one of the capacitors. A semiconductor substrate of p-type silicon is covered with a thin layer of insulating silicon oxide which insulates the Si substrate from the metal electrode. When a positive voltage is applied between the electrode and the Si substrate, the minority carriers (holes in p-type Si) will be repelled

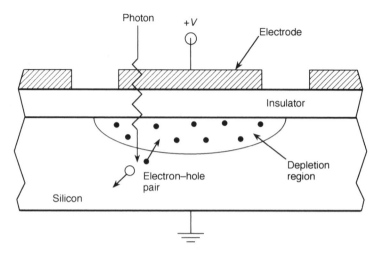

Fig. 5.20. *Simplified sketch of an MOS capacitor*

from the interface between semiconductor and insulator, creating a region free of mobile carriers directly underneath the electrode. This region is known as the depletion region and has a thickness of a few micrometres. The metal electrodes (usually made of polycrystalline silicon) are transparent for wavelengths larger than about 400 nm.

If an incident photon has an energy larger than the bandgap in Si, it can create an electron–hole pair in the semiconductor. When this creation occurs in or near the depletion region, the photon generated electron is attracted towards the potential well which is formed under the positive charged electrode. In this way, a charge packet is formed consisting of photon-electrons which were created in the vicinity of a specific electrode.

Figure 5.21 shows how the charge packets are transferred from one electrode to the next by proper clocking of the potentials of the electrodes. In a device having a planar oxide and uniform substrate doping, at least three phases are required for unidirectional charge transfer, i.e. a barrier is maintained behind the charge packet while a deeper well is formed in front of the packet. The clocking diagrams are shown in Figure 5.21(b). At $t = t_1$, charge resides in the wells under the ϕ_1 electrodes. At $t = t_2$, the potential on ϕ_2 is made positive forming wells under the ϕ_2 electrodes. Charge will then flow from the ϕ_1 wells into the ϕ_2 wells. At $t = t_3$, the potential on the ϕ_1 electrodes is reduced to a low value, and the remaining charge in the ϕ_1 wells will be pushed into the ϕ_2 wells. This sequence repeats with the result that the charge configuration moves from one cell to the next every clock period.

To allow the device to be clocked with two phases, potential barriers between electrodes must be built in. This is done either by forming alternate thin and thick oxide insulators, the so-called stepped oxide barrier, or the implanted barrier technique by non-uniform doping of the substrate.

There are two different ways in which CCDs are organized when applied as an imaging sensor. In the following description standard TV rates are assumed, i.e. 1/25 s, European standard (CCIR) or 1/30 s, American standard (RS-170) frame time. We also assume two-phase clocking of the electrodes. Figure 5.22(a) shows the organization of the

Fig. 5.21. *Three-phase CCD: (a) cross-sectional view showing input section, transfer section, and output section; a primitive electrode structure having unprotected gaps is shown for simplicity; (b) surface-potential profile showing potential wells under the ϕ_1 electrodes; (c) surface-potential profiles showing progression of charge transfer during one clock period; (d) clocking waveforms used to drive the CCD during transfer. (From Barbe 1975. Reproduced by permission of IEEE)*

so-called frame-transfer structure. This sensor is divided into two identical areas, the image section and a masked storage section. A TV-frame is divided into two fields A and B, see Section 5.8. Field A is formed by collecting photoelectrons under the odd rows of electrodes for $1/50$ s ($1/60$ s for RS-170). This charge configuration is shifted into the shielded storage register in a time that is short (several MHz) compared with the integration (exposure) time. Field A is then read out a line at a time while field B is being formed by collecting photoelectrons under the even electrodes.

Figure 5.22(b) shows the organization of the so-called interline transfer structure. Here the shielded vertical readout registers are interdigitated with the photosensitive column. Potential wells are formed in the photosensitive regions by applying voltages to the vertical polycrystalline silicon (polysilicon) stripes. The horizontal polysilicon stripes are used to clock the vertical shielded register. Because the integrating cells and shift-out cells are separate, the effective integration time for both fields A and B is $1/25$ s. The operation is as follows. After collecting photoelectrons in field A for $1/25$ s, the charge configuration is shifted into the shielded registers and down, a line at a time, into the horizontal output

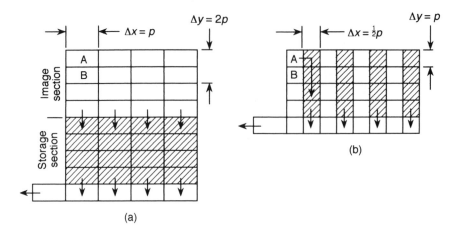

Fig. 5.22. *Schematic diagrams showing the (a) frame-transfer and (b) inter-line transfer array organizations*

register. When field A has been completely read out (1/50 s), field B is shifted into the shielded registers and out. Note that the effective integration time for the interline transfer structure is twice that of the frame-transfer structure because the integration in the interline transfer structure is performed in sites separate from the transport registers.

5.6.2 Responsivity

A very important parameter for an imaging sensor used for optical metrology is its responsivity R, particularly the responsivity as a function of the spatial frequency $R(f)$ of the imaged scene. This can be formulated as

$$R(f) = R(0) \prod_i (MTF)_i \tag{5.44}$$

where $R(0)$ is the responsivity at zero spatial frequency and the other term is the product of all of the modulation transfer function (MTF, see Section 4.6.2) factors that affect the frequency response of the chip. These factors are: (1) the loss of frequency response due to the geometry of the integrating cell (MTF_{integ}), (2) the loss of frequency response due to transfer inefficiency ($MTF_{transfer}$), and (3) the loss of frequency response due to the diffusion of charge between photon absorption and photoelectron collection (MTF_{diff}). We here will consider three chip designs: The front-illuminated interline transfer CCD (FIIT), the front illuminated frame transfer CCD (FIFT) and the back-illuminated frame transfer CCD (BIFT). For FIIT approximately one-half of the chip area is photosensitive because the other half is occupied by the vertical transport registers. When used in standard TV interlaced mode, however, the integration time for FIFT and BIFT is half of that for FIIT. The efficiency with which photons are absorbed and the resulting photo-electrons are collected in the integrating cells are twice as high for BIFT than for both FIIT and FIFT when averaged over the 0.4–1.0 µm wavelength band. In conclusion therefore, R(0) is twice as high for BIFT than for FIIT and FIFT.

In modern CCD chips, $MTF_{transfer}$ and MTF_{diff} have negligible effects compared to MTF_{integ}. The latter is, however, a fundamental effect on the chip responsivity and is related to the finite size of the integrating cells.

To calculate MTF_{integ} it is sufficient to find the response to a sinusoidal grating of frequency f

$$H_1 = H_0[1 + m\cos 2\pi fx] \tag{5.45}$$

When sensed by an array of cells of width Δx and inter-cell distance p, the output charge

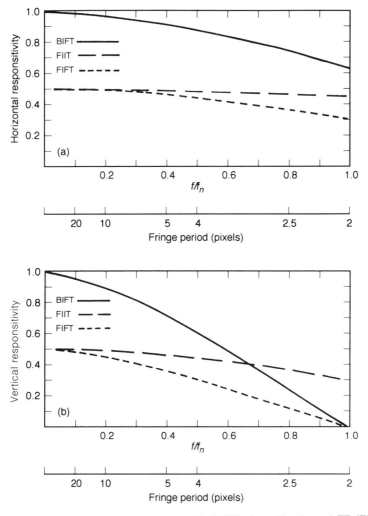

Fig. 5.23. *Responsivity for back-illuminated FT (BIFT), front-illuminated FT (FIFT) and IT (FIIT) arrays: (a) versus normalized horizontal spatial frequency and fringe period; (b) versus normalized vertical spatial frequency and fringe period. The relations between Δx, Δy and p are as in Figure 5.22, i.e. standard video signal transfer is assumed*

pattern becomes

$$H_2 = \frac{1}{p}\int_{x-\Delta x/2}^{x+\Delta x/2} H_1 \, dx = \frac{H_0}{p}\int_{x-\Delta x/2}^{x+\Delta x/2} [1 + m\cos 2\pi fx]\, dx$$

$$= \frac{H_0\Delta x}{p}\left[1 + m\frac{\sin\pi\Delta x}{\pi f\Delta x}\cos 2\pi fx\right] \tag{5.46}$$

Therefore the MTF of the integrating process is (cf. Section 4.6.2, Equation (4.54))

$$MTF_{integ} = \left|\frac{\sin\pi f\Delta x}{\pi f\Delta x}\right| \tag{5.47}$$

By introducing the so-called Nyquist frequency $f_n = 1/2p$, this can be written

$$MTF_{integ} = \left|\frac{\sin\dfrac{\pi}{2}\dfrac{f}{f_n}\dfrac{\Delta x}{p}}{\dfrac{\pi}{2}\dfrac{f}{f_n}\dfrac{\Delta x}{p}}\right| \tag{5.48}$$

Figure 5.23 shows the responsivity as a function of spatial frequency for FIIT, FIFT and BIFT in the horizontal and vertical directions.

5.7 SAMPLING

5.7.1 Ideal sampling

Consider a one-dimensional function $f(x)$ which might represent e.g. the irradiance distribution along a TV-line on a CCD camera. To sample this function means to find the values of f at regular intervals, i.e. $f(np)$ where $n = 0, 1, 2, \ldots$ and p is a constant called the sampling period (Goodman 1968). This is equivalent to multiplying f by a comb function (Equation (B.16), Appendix B.2) to get the sampled function f_s,

$$f_s = \mathrm{comb}\left(\frac{x}{p}\right)f(x) \tag{5.49}$$

The spectrum F_s of f_s is given by its Fourier transform

$$F_s(f_x) = \mathscr{F}\left\{\mathrm{comb}\left(\frac{x}{p}\right)f(x)\right\} = \mathscr{F}\left\{\mathrm{comb}\left(\frac{x}{p}\right)\right\}\otimes F(f_x) \tag{5.50}$$

where the last equality follows from the convolution theorem and $F(f_x)$ is the spectrum of $f(x)$. Now we have that

$$\mathscr{F}\left\{\mathrm{comb}\left(\frac{x}{p}\right)\right\} = \mathrm{pcomb}(pf_x) = \sum_{n=-\infty}^{\infty}\delta\left(f_x - \frac{n}{p}\right) \tag{5.51}$$

It follows that the spectrum of the sampled function is given by

$$F_s(f_x) = \sum_{n=-\infty}^{\infty} F\left(f_x - \frac{n}{p}\right) \tag{5.52}$$

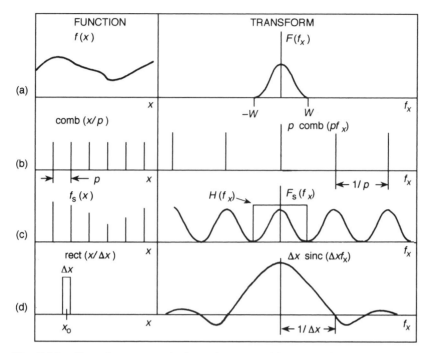

Fig. 5.24. *Sampling process in the x-domain and the spatial frequency domain*

Evidently the spectrum of f_s can be found simply by erecting the spectrum of f about each point n/p along the f_x-axis as shown in Figure 5.24c).

Now assume (as in Figure 5.24(a)) that the spectrum F of f vanishes outside some interval $[-W, W]$. A function whose transform has this property for any finite value of W is called a band-limited function. From Figure 5.24(c) we see that if

$$\frac{1}{p} \geqslant 2W \tag{5.53}$$

the spectra $F(f_x - n/p)$ constituting the spectrum of f_s do not overlap. If inequality (5.53) is fulfilled, we therefore can separate the $n = 0$ term of F_s from all the other terms by multiplying it by a filter function

$$H(f_x) = \mathrm{rect}\left(\frac{f_x}{2W}\right) \tag{5.54}$$

We see therefore that F is recovered from F_s in that

$$F_s(f_x)\mathrm{rect}\left(\frac{f_x}{2W}\right) \equiv F(f_x) \tag{5.55}$$

The inverse Fourier transform of Equation (5.55) yields

$$\mathscr{F}^{-1}\{F(f_x)\} = f(x) = \mathscr{F}^{-1}\left\{F_s(f_x)\mathrm{rect}\left(\frac{f_x}{2W}\right)\right\} = \mathscr{F}^{-1}\{F_s(f_x)\} \otimes \mathscr{F}^{-1}\left\{\left(\mathrm{rect}\left(\frac{f_x}{2W}\right)\right)\right\}$$

$$= f_s(x) \otimes 2W\,\mathrm{sinc}(2Wx) = f(x)\,\mathrm{comb}\left(\frac{x}{p}\right) \otimes 2W\,\mathrm{sinc}(2Wx) \tag{5.56}$$

Noting that

$$f(x)\,\mathrm{comb}\left(\frac{x}{p}\right) = p\sum_{n=-\infty}^{\infty} f(np)\delta(x - np) \tag{5.57}$$

Equation (5.56) becomes

$$f(x) = 2pW\sum_{n=-\infty}^{\infty} f(np)\,\mathrm{sinc}[2W(x - np)] \tag{5.58}$$

Finally, when the sampling interval p is taken to have its maximum allowable value $1/2W$, we have that

$$f(x) = \sum_{n=-\infty}^{\infty} f\left(\frac{n}{2W}\right)\mathrm{sinc}\left[2W\left(x - \frac{n}{2W}\right)\right] \tag{5.59}$$

Equation (5.59) represents a fundamental result which we refer to as the Whittaker–Shannon sampling theorem. It implies that exact recovery of a bandlimited function can be achieved from an appropriately spaced array of its sampled values, the recovery is accomplished by injecting, at each sample point, an interpolation function consisting of a sinc function.

It should be noted that other choices of the filter function $H(f_x)$ than that given in Equation (5.54) is possible as long as $H(f_x)$ passes the $n = 0$ term of F_s and excludes other terms. In fact it is a multitude of choices which will result in alternative sampling theorems.

5.7.2 Non-ideal sampling

The sampling of a function by discrete points is an idealized situation. In reality the sampling pulses always have finite width.

Consider Figure 5.24d) with such a sampling pulse of width Δx centred at x_0. If this pulse represents the cell in a CCD chip, the irradiance $f(x)$ will be integrated over this cell. The charge at x_0 will therefore be given by

$$f_i(x_0) = \int_{x_0 - \Delta x/2}^{x_0 + \Delta x/2} f(x)\,\mathrm{d}x \tag{5.60}$$

If we introduce the rectangle function (Equation (B.13), Appendix B.2), this integral can be written as

$$f_i(x_o) = \int_{-\infty}^{\infty} \mathrm{rect}\left(\frac{x - x_o}{\Delta x}\right) f(x)\,dx \tag{5.61}$$

Since $\mathrm{rect}(-x) = \mathrm{rect}(x)$ (it is symmetric), we have

$$f_i(x) = \int_{-\infty}^{\infty} \mathrm{rect}\left(\frac{x - \xi}{\Delta x}\right) f(\xi)\,d\xi \tag{5.62}$$

where we also have changed variables. Equation (5.62) is recognized as a convolution integral, i.e.

$$f_i(x) = \mathrm{rect}\left(\frac{x}{\Delta x}\right) \otimes f(x) \tag{5.63}$$

This function is then sampled in the same way as for ideal sampling

$$f_{sN} = f_i(x)\,\mathrm{comb}\left(\frac{x}{p}\right) = \left[\mathrm{rect}\left(\frac{x}{\Delta x}\right) \otimes f(x)\right]\mathrm{comb}\left(\frac{x}{p}\right) \tag{5.64}$$

The spectrum now becomes

$$F_{sN} = \mathscr{F}\left\{\mathrm{rect}\left(\frac{x}{\Delta x}\right) \otimes f(x)\right\} \otimes \mathscr{F}\left\{\mathrm{comb}\left(\frac{x}{p}\right)\right\}$$

$$= \Delta x\,\mathrm{sinc}(\Delta x f_x) F(f_x) \otimes \mathrm{pcomb}(p f_x) = \sum_{n=-\infty}^{\infty} F_N\left(f_x - \frac{n}{p}\right) \tag{5.65}$$

where

$$F_N = \Delta x\,\mathrm{sinc}(\Delta x f_x) F(f_x) \tag{5.66}$$

Apart from $F(f_x)$ being multiplied by a sinc function, this is the same result as for ideal sampling. This does not matter so much as long as f_x is well below $1/\Delta x$. (The first zero of the sinc-function, see Figure 5.24(d)).

5.7.3 Aliasing

If inequality (5.53) is not fulfilled, the repeated spectra F_N will overlap each other as seen in Figure 5.25. Since natural scenes are not band limited, the spectra will always overlap unless $F(f_x)$ is prefiltered. Overlapping of the spectra causes frequencies higher than the Nyquist limit ($f_n = 2/p$) to appear in the passband ($-f_n \leqslant f_x \leqslant f_n$) as lower-frequency components – thus the term 'aliasing'. Thus, for example the frequency $1.5 f_n$ in F_N for $n = 1$ would give a response in F_N for $n = 0$ at $0.5 f_n$.

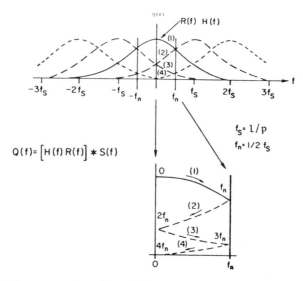

$$Q(f) = \left[H(f)\,R(f)\right] * S(f)$$

Fig. 5.25. *Spatial frequency domain analysis of the sampling process. Branch (1) is the desired response, and branches (2), (3) and (4) represent response to $f > f_n$ aliased into the passband. (From Barbe, D.F. and Campana, S.B., (1976) Aliasing and MTF effects in photosensor arrays, in P.G. Jespers, F. v.de Wiele and M.H. White (eds), Solid State Imaging, Noordhoff, Leyden. Reproduced by permission of D.F. Barbe)*

Fig. 5.26. *Bar-patern imagery produced by the 100 × 100 element CCD array. The fundamental frequencies are given relative to the Nyquist frequency, f_n. (From Barbe, D.F. and Campana, S.B., (1976) Aliasing and MTF effects in photosensor arrays, in P.G. Jespers, F. v.de Wiele and M.H. White (eds), Solid State Imaging, Noordhoff, Leyden. Reproduced by permission of D.F. Barbe)*

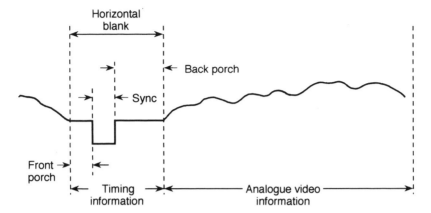

Fig. 5.27.   *The analog video signal with timing information*

An example of aliasing is shown in Figure 5.26. Here vertical bars of different spacings are imaged onto a 100×100 element interline transfer CCD chip. The Nyquist frequency in this case was 12.3 cycles/mm. Thus only the top row of Figure 5.26 represents the true imagery. The remaining six views are moirè patterns produced by the interaction of the CCD structure and the bars of the test chart.

5.8 SIGNAL TRANSFER

Most electronic cameras are equipped with a video output signal. This is an analog signal containing the image data (Grob 1984). To guide the scanning beam of the TV-monitor, this video signal also contains some timing information, see Figure 5.27. The timing information is transmitted between each horizontal scan line and is called the horizontal blanking period. In this period the electron beam intensity is set to zero to avoid disturbing the picture during flyback to the next TV line. The horizontal blanking period consists of a horizontal (hsync) pulse which is used for synchronization and a back porch. In the AD converter of the frame grabber (see Section 10.1), this porch is used for adjusting the grey level to a known value (black). All video signals use interlaced raster technique. This means that the horizontal scan lines of a TV frame are divided into two fields. The even field consists of all the even-numbered lines in the frame, starting with line zero. The odd fields consists of the odd-numbered scan lines. This is done to avoid flicker on the TV-screen. The even and odd fields are separated by a vertical blanking period.

A complete scanning pattern is shown in Figure 5.28, where the corresponding horizontal and vertical sawtooth waveforms illustrate odd-line interlaced scanning. A total of 21 lines in the frame is used for simplicity, instead of 525 (American standard). The 21 lines are interlaced with two fields per frame. Of the 10.5 lines in a field, we can assume that 1 line is scanned during vertical retrace to have a convenient vertical flyback time. So 9.5 lines are scanned during vertical trace in each field. Therefore in the entire frame, 19 lines are scanned during vertical trace, so 2 lines are lost in the vertical retrace lines. Starting at point A in the upper left corner, the beam scans the first line from left to right and retraces

Fig. 5.28. *A sample scanning pattern for 21 interlaced lines per frame and $10\frac{1}{2}$ lines per field. The corresponding H and V sawtooth deflection waveforms are shown below pattern. Starting at point A, the scanning motion continues through B, C and D and back to A again. ((From Groh, B. (1984) Basic Television and Video Systems. (5th edn), McGraw-Hill, N.Y. Reproduced by permission of McGraw-Hill Inc., N.Y)*

to the left to begin scanning the third line in the frame. Then the beam scans all succeeding odd lines until it reaches point B at the bottom, when vertical flyback begins. Note that this vertical retrace begins in the middle of a horizontal line. During this vertical retrace the scanning beam is brought to point C, which is separated from A by exactly one half-line so that the scanning of the second field can begin. In commercial TV broadcasting, some of the 'lost' lines in vertical retrace are digitally encoded to carry data for reproduction of full pages of alphanumeric characters for videotext. This requires a decoder at the receiver to gate out the specific lines and process the digital signal.

Unfortunately, the video signal has several standards. Analogue input comes in three different forms, essentially distinguished by the corresponding 8-bit conversion frequency: slow-scan, standard video and high definition video (HDTV). Table 5.5 shows the different frequencies, frame grabber sizes and grey value resolutions.

The standard video norms for black and white are RS-170 (used in North and South America and Japan) and CCIR (used in Europe). A detailed overview of these systems is found in Table 5.6.

Table 5.5 *Three different television standard video systems distinguished in acquisition speed, frame size and greyvalue resolution*

Pixel frequency	Slow scan <5 MHz	Standard video 7.5–15 MHz	HDTV >15 MHz
Frame size in pixels	1024 × 1024 4096 × 4096	256 × 256 512 × 512 780 × 540	1280 × 1024
Grey-value resolution	6–16 bit	6–10 bit	6–8 bit

Table 5.6 *Comparison of the black and white video systems RS-170 and CCIR*

	RS-170	CCIR
Frame rate/field rate	30/60 Hz	25/50 Hz
Number of lines	525	625
Number of active lines	480	576
Field time	16 2/3 ms	20 ms
Time per line	63.49 μs	64 μs
Active line period	52.5 μs	52 μs
Nominal video bandwidth	4.5 MHz	5.5 MHz
Resolution	472	572
Aspect ratio	4:3	4:3

Standard video uses interlacing to avoid flicker in human perception of video images. This is not demanded in machine vision and video metrology, but will in many cases be a drawback, especially when monitoring high-speed phenomena. When using non-interlaced video, the resolution in the vertical direction of the frame transfer CCD cameras is also doubled, see Section 5.6.2. Specialized camera manufacturers such as EG & G, Fairchild and Dalsa offer matrix cameras with capabilities of up to 6000 × 6000 pixels for non-standard video transfer. The scan rate is driven by an external clock and can be selected by the operator. Such cameras are of course superior to standard video cameras with regard to resolution, but the amount of data and therefore the processing time increases dramatically, from $512 \times 512 \times 8 = 200$ kbyte to 288 Mbyte for a single image.

As mentioned in Section 10.1, for frame grabbers receiving composite video, line jitter is a problem especially when making measurements with sub-pixel accuracy. This problem is avoided when the signal is transferred digitally from the camera to the frame grabber and not via an analogue video signal. This is achieved by placing the A/D converter inside the camera and transferring the signal via a data cable. Such cameras are manufactured by Kodak (Videk Megaplus) and Cohu. Most frame grabbers today accept such signals.

PROBLEMS

5.1 Assume the Sun to be a 6000 K blackbody source and that its diameter subtends an angle $\alpha = 9.3$ mrad at the Earth.

(a) Find the wavelength λ_m corresponding to the maximum solar spectral radiant exitance M_λ.
(b) Find M_λ of the Sun's surface at this wavelength.
(c) What is the spectral radiance L?

The area of the solar disc dA is given by

$$dA = \pi \alpha^2 s^2 / 4$$

where s is the Earth–Sun distance.
(d) Calculate the radiant flux Φ on an area of 1 m^2 (i.e. the irradiance) on the top of the earth's atmosphere at the wavelength λ_m.

5.2 A Lambert source S with radiance $L_e = 10^6$ W/m^2, 1 cm in diameter, is located 45 cm from a thin lens of 30 cm focal length and 10 cm diameter, see Figure P5.1. A detector is located 120 cm to the right of the lens. It consists of two masks M_1 and M_2 each with 0.1 mm diameter pinholes on axis 1 cm apart, followed closely by a large photocell PC with a sensitivity of 10^3 A/W. Neglect losses in the lens and find the value of the photocurrent coming from the cell.

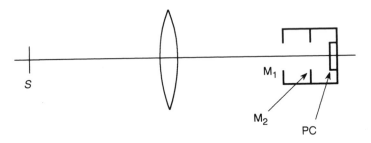

Fig. P5.1

5.3 The responsivity of the integrating process, cf. Equations (5.44), (5.48), we write as

$$R(f) = \frac{\Delta x}{p} \frac{\sin\left(\dfrac{\pi}{2} \dfrac{f}{f_n} \dfrac{\Delta x}{p}\right)}{\dfrac{\pi}{2} \dfrac{f}{f_n} \dfrac{\Delta x}{p}} = r \frac{\sin\left(\dfrac{\pi}{2} qr\right)}{\dfrac{\pi}{2} qr} = R(q, r)$$

where $q = f/f_n$, $r = \Delta x/p$. Assume that r can be varied between 0 and 1. The lowest possible relative frequency q for which $R(q, r) = R(q, 1)$, we denote q_e.

(a) Find q_e expressed by r. Can $R(q, r) > R(q, 1)$ for $q < 1$?
(b) Find q_e for $r = 0.5$.

5.4 Light sources producing uniform beams is an idealization. In reality the intensity varies across the transverse plane. A particularly important transverse pattern is the Gaussian distribution, radiated by most gas lasers and some specially designed laser diodes and also single-mode fibres. The Gaussian field distribution is given as

$$U = U_o \exp\left[-\frac{r^2}{w^2}\right]$$

Given that

$$\int_{-\infty}^{\infty} e^{-u^2} du = \sqrt{\pi}$$

(a) Calculate the one-dimensional Fraunhofer diffraction pattern of this distribution. An accepted definition of the radius of the spot size of this beam is the distance at which the intensity has dropped to $1/e^2 = 0.135$.
(b) Calculate the spot size in the focal plane of a lens (with diameter $D > w$) of focal length f. Compare the resulting pattern with the Airy pattern.

5.5 Consider a laser oscillating in three resonator modes of finite width Δv_h. The translated spectral distribution function $S(v)$ then can be written as

$$S(v) = G(v) \otimes \tfrac{1}{2}[\delta(v) + \tfrac{1}{2}\delta(v - \Delta v) + \tfrac{1}{2}\delta(v + \Delta v)]$$

where each mode has a Gaussian distribution

$$G(v) = \sqrt{\left(\frac{2}{\pi}\right)}\frac{1}{\Delta v_h}\exp\left[-2\left(\frac{v}{\Delta v_h}\right)^2\right]$$

and where Δv is the resonator frequency spacing.
   Calculate the temporal degree of coherence $\gamma(\tau)$. Assume $\Delta v = 3\Delta v_h$ and compare the result with that found in Section 5.3.9.

5.6 To look more closely into the sampling process, consider the simple input signal

$$f(x) = \tfrac{1}{2}(1 + \cos 2\pi f_o x)$$

(a) What is the bandwidth $2W$ of this signal?
This signal is ideally sampled at a sampling interval of p (i.e a sampling frequency $f_s = 1/p$).
(b) Find the spectrum of the sampled signal. Assume that we apply a filter function $H(f_x) = \text{rect}(f_x/2W)$ with the value of W found in (a).
   Find the resulting spectrum and the output signal in the following cases:
(i) $f_s > 2f_o$, (ii) $f_s = 1.5f_o$, (iii) $f_s = 0.4f_o$.

6 Holography

6.1 THE HOLOGRAHIC PROCESS

Figure 6.1(a) shows a typical holography set-up (Gabor 1949; Leith and Upatnieks 1962). Here the light beam from a laser is split in two by means of a beamsplitter. One of the partial waves is directed onto the object by a mirror and spread to illuminate the whole object by means of a microscope objective. The object scatters the light in all directions, and some of it impinges onto the hologram plate. This wave is called the object wave. The other partial wave is reflected directly onto the hologram plate. This wave is called the reference wave. In the figure this wave is collimated by means of a microscope objective and a lens. This is not essential, but it is important that the reference wave constitute a uniform illumination of the hologram plate. The hologram plate must be a light-sensitive medium, most commonly a silver halide film plate with high resolution. We now consider the mathematical description of this process in more detail. For more comprehensive treatments, see Collier *et al.* (1971), Smith (1969), Caulfield (1979) and Hariharan (1984).

Let the object and reference waves in the plane of the hologram be described by the field amplitudes u_o and u respectively. These two waves will interfere, resulting in an intensity distribution in the hologram plane given by

$$I = |u + u_o|^2 = |u|^2 + |u_o|^2 + u_o^* u + u_o u^* \tag{6.1}$$

We now expose the film plate to this intensity distribution until it gets a suitable blackening. Then it is removed from the plate-holder and developed. We now have a hologram, and the process so far is called a hologram recording.

This hologram will have an amplitude transmittance t which is proportional to the intensity distribution given by Equation (6.1). This means

$$t = \alpha I \tag{6.2}$$

We then replace the hologram back in the holder in the same position as in the recording. We block the object wave and illuminate the hologram with the reference wave which now is termed the reconstruction wave (see Fig. 6.1(b)). The amplitude distribution u_a just behind the hologram then becomes equal to the field amplitude of the reconstruction wave multiplied by the amplitude transmittance of the hologram, i.e.

$$u_a = t \cdot u = \alpha [|u|^2 + |u_o|^2]u + \alpha(uu)u_o^* + \alpha|u|^2 u_o \tag{6.3}$$

As mentioned above, the reference (reconstruction) wave is a wave of uniform intensity. The quantity $|u|^2$ is therefore a constant and the last term of Equation (6.3) thus becomes (apart from a constant) identical to the original object wave u_o. We therefore have been able to reconstruct the object wave, maintaining its original phase and amplitude distribution. The consequence is that, by looking through the hologram in the direction of

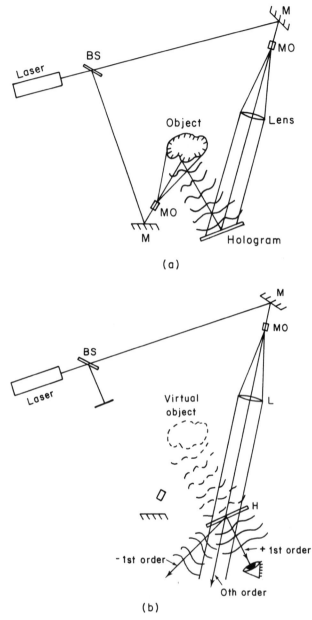

Fig. 6.1. *(a) Example of a holography set-up. BS = beamsplitter, M = mirrors, MO = microscope objectives. (b) Reconstruction of the hologram*

the object, we will observe the object in its three-dimensional nature even though the physical object has been removed. Therefore this reconstructed wave is also called the virtual wave.

The two other terms of Equation (6.3) represent waves propagating in the directions indicated in Figure 6.1(b). In fact, a hologram can be regarded as a very complicated

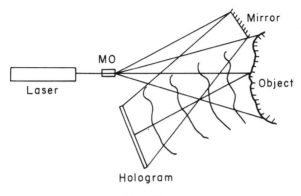

Fig. 6.2.

grating where the first term of Equation (6.3) represents the zeroth order and the second and third terms represent the $\pm$ 1st side orders diffracted from the hologram. If we could use u^*, the conjugate of u, as the reconstruction wave, we see that the second term of Equation (6.3) would have become proportional to $|u|^2 u_o^*$, i.e. the conjugate of the object wave would have been reconstructed. The physical meaning of this deserves some explanation. Complex conjugation of a field amplitude means changing the sign of its phase. It thus gives a wave field returning back its own path. u_o^* therefore represents a wave propagating from the hologram back to the object forming an image of the object. It is therefore termed the real wave. To reconstruct the hologram with u^* in the case of a pure plane wave as the reference (reconstruction) wave, the reconstruction wave can be reflected back through the hologram by means of a plane mirror. An easier way, which also applies for a general reference (reconstruction) wave, is to turn the hologram 180° around the vertical axis. By placing a screen in the real wave, we then can observe the image of the object on the screen.

In Figure 6.2 another possible realization of a holography set-up is sketched. Here the expanded laser beam is wavefront-divided by means of a mirror which reflects the reference wave onto the hologram. This set-up is normally more stable than that in Figure 6.1 since fewer components are involved.

6.2 AN ALTERNATIVE DESCRIPTION

An alternative and more physical description of the holographic process has already been touched on in Section 4.3.1. Let the point source P in Figure 4.7(a) represent the light from a point on the object, and let the plane wave represent the reference wave. The resulting zone plate pattern is recorded on a film. In Fig. 4.7(b) this developed film (the hologram) is illuminated by a plane wave (the reconstruction wave). When viewed through the film, the diffracted, diverging spherical wave looks as if it is coming from P. This argument can be repeated for all points on the object and give us the virtual reconstructed object wave. The spherical wave converging to point P' represents the real wave.

The circular zone plate is therefore also termed a unit hologram. In the general case when the object- and reference waves are not normally incident on the hologram, the pattern changes from circular to elliptical zone plate patterns, and the diffracted virtual and real waves propagate in different directions in the reconstruction process.

6.3   DIFFRACTION EFFICIENCY. THE PHASE HOLOGRAM

Assume the object- and reference waves to be described by

$$u_o = U_o e^{i\phi_o} \tag{6.4a}$$

and

$$u = U e^{i\phi} \tag{6.4b}$$

respectively. The resulting amplitude transmittance then becomes

$$t = \alpha[U^2 + U_o^2 + UU_o e^{i(\phi-\phi_o)} + UU_o e^{-i(\phi-\phi_o)}]$$
$$= \alpha(I + I_0)[1 + V\cos(\phi - \phi_0)] \tag{6.5}$$

which can be written as

$$t = t_b\left[1 + \frac{V}{2}e^{i(\phi-\phi_o)} + \frac{V}{2}e^{-i(\phi-\phi_o)}\right] \tag{6.6}$$

where $I = U^2$, $I_0 = U_0^2$ and where we have introduced the visibility V (see eq. (3.29)) and the bias transmittance $t_b = \alpha(I + I_o)$. Since the transmittance t never can exceed unity and $0 \leqslant V \leqslant 1$, we see from Equation (6.5) that $t_b \leqslant 1/2$.

The reconstructed object wave u_r is found by multiplying the last term of Equation (6.6) by the reconstruction wave u:

$$u_r = t_b \frac{V}{2} U e^{i\phi_o} \tag{6.7}$$

and the intensity

$$I_r = |u_r|^2 = \tfrac{1}{4}U^2 t_b^2 V^2 \tag{6.8}$$

The diffraction efficiency η of such a hologram we define as the ratio of the intensities of the reconstructed wave and the reconstruction wave, i.e.

$$\eta = I_r/I = \tfrac{1}{4}t_b^2 V^2 \tag{6.9}$$

From this expression we see that the diffraction efficiency is proportional to the square of the visibility. η therefore reaches its maximum when $V = 1$, i.e. when $I_o = I$, which means that the diffraction efficiency is highest when the object and reference waves are of equal intensity.

Maximum possible diffraction efficiency is obtained for $V = 1$ and $t_b = \tfrac{1}{2}$, which gives

$$\eta_{max} = \tfrac{1}{16} = 6.25\%$$

This type of hologram is called an amplitude hologram because its transmittance is a pure amplitude variation. A hologram with a pure phase transmittance is called a phase hologram. Such holograms can be produced in different ways. A commonly applied method consists of bleaching the exposed silver grains in the film emulsion of a standard amplitude hologram. The recorded amplitude variation then changes to a corresponding variation in emulsion thickness. The transmittance t_p of a phase hologram formed by bleaching of an amplitude hologram can be written as

$$t_p = e^{iM\cos(\phi_0 - \phi)} = \sum_{n=-\infty}^{\infty} i^n J_n(M)e^{in(\phi_0 - \phi)} \tag{6.10}$$

where J_n is the nth-order Bessel function. Here M is the amplitude of the phase delay. From this expression we see that a sinusoidal phase grating will diffract light into n orders in contrast to a sinusoidal amplitude grating which has only $\pm$ 1st orders. The amplitude of the first-order reconstructed object wave is found by multiplying Equation (6.10) by the reconstruction wave u for $n = 1$, i.e.

$$u_r = J_1(M)Ue^{i\phi_0} \tag{6.11}$$

and the intensity

$$I_r = U^2 J_1^2(M) \tag{6.12}$$

The diffraction efficiency becomes

$$\eta_p = I_r/I = J_1^2(M) \tag{6.13}$$

Since $J_{1\max}(M) = 0.582$ for $M = 1.8$, the maximum possible diffraction efficiency of a phase hologram is

$$\eta_{p,\max} = 0.339 = 34\%$$

For converting an amplitude hologram recorded in silver halide emulsion into a phase hologram, there exist a number of formulas. Here we present a relatively simple process due to Hariharan *et al.* (1971). The dichromate–iodide bleach consists of

Stock solution A

potassium dichromate	8 g
conc. sulphuric acid	10 ml
Distilled water to make	1 litre

Stock solution B

potassium iodide	2 g
distilled water to make	1 litre

Mix 1 part of A, 1 part of B and 18 parts distilled water just before use.
The processing schedule is as follows:

(1) Expose the hologram to a density (after normal development) of 2.0 to 2.5.

(2) Develop in D-19 for 5 minutes.

(3) Rinse in distilled water for 3 minutes (two changes).

(4) Bleach for 5 minutes (bleach formula given above).

(5) Rinse in distilled water for 6 minutes (four changes, to the last change add 2 ml Kodak Photo-Flo per litre).

(6) Dry at 20°C and 40–60% relative humidity.

6.4 VOLUME HOLOGRAMS

Up to now we have regarded the hologram film emulsion as having negligible thickness.

For emulsions of non-negligible thickness, however, volume effects, hitherto not considered, must be taken into account. For example, a thick phase hologram can reach a theoretical diffraction efficiency of 100 per cent.

Consider Figure 6.3(a) where two plane waves are symmetrically incident at the angles $\theta/2$ to the normal on a thick emulsion. These waves will form interference planes parallel to the yz-plane with spacings (cf. eq. (3.21)).

$$d = \frac{\lambda}{2 \sin (\theta/2)} \tag{6.14}$$

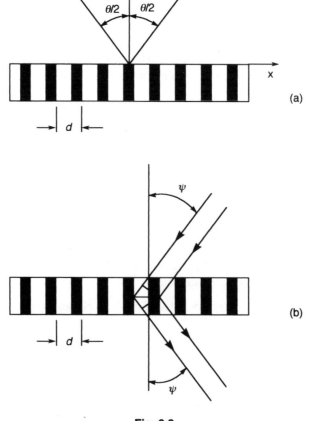

Fig. 6.3

After development of this hologram, the exposed silver grains along these interference planes will form silver layers that can be regarded as partially reflecting plane mirrors. In Figure 6.3(b) this hologram is reconstructed with a plane wave incident at an angle ψ. This wave will be reflected on each 'mirror' at an angle ψ.

To obtain maximum intensity of the reflected, reconstructed wave, the path length difference between light reflected from successive planes must be equal to λ. From the triangles in Figure 6.3(b) this gives

$$2d \sin \psi = \lambda \tag{6.15}$$

which, by substitution of Equation (6.14), gives

$$\sin \psi = \sin \theta/2 \tag{6.16}$$

i.e. the angles of incidence of the reconstruction and reference waves must be equal. It can be shown that for a thick hologram, the intensity of the reconstructed wave will decrease rapidly as ψ deviates from $\theta/2$. This is referred to as the Bragg effect and Equation (6.16) is termed the Bragg law.

A special type of volume hologram, called a reflection hologram, is obtained by sending the object and reference waves from opposite sides of the emulsion, as shown in Figure 6.4(a). Then $\theta = 180°$ and the stratified layers of metallic silver of the developed hologram run nearly parallel to the surface of the emulsion with a spacing equal to $\lambda/2$ (see

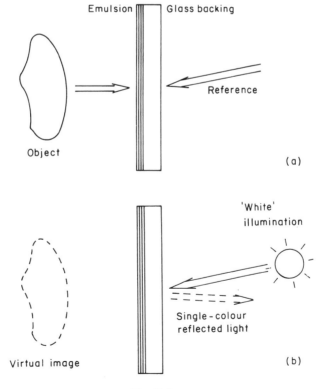

Fig. 6.4.

Equation (6.14)). Owing to the Bragg condition, the reconstruction wave must be a duplication of the reference wave with the same wavelength, i.e. the hologram acts as a colour filter in reflection. Therefore a reflection hologram can be reconstructed in white light giving a reconstructed wave of the same wavelength as in the recording (see Figure 6.4(b)). In practice the wavelength of the reflected light is shorter than that of the exposing light, the reason being that the emulsion shrinks during the development process and the silver layers become more closely spaced.

6.5 STABILITY REQUIREMENTS

In the description of the holographic recording process we assumed the spatial phases of both the object- and reference waves to be time independent during exposure. It is clear, however, that relative movements between the different optical components (like mirrors, beamsplitters, the hologram, etc.) in the hologram set-up will introduce such phase changes. If, for instance, a mirror makes vibrations of amplitude greater than $\lambda/4$ during the exposure time, adjacent dark and bright interference fringes interchange their positions randomly, which can lead to a uniform blackening of the hologram film and therefore ruin the experiment. The exposure time using a 5 mW H-Ne laser is typically of the order of seconds. This poses stringent requirements on the stability of the set-up. Therefore the standard methods of holography are normally performed on vibration-isolated heavy tables with the optical components mounted in massive holders. There are, however, special techniques by which unwanted movements can to a certain extent be compensated for or subtracted from. Some of these will be treated in Section 6.8. Thus, successful holographic experiments executed on the factory floor using continuous wave lasers have been reported.

By using pulse lasers, exposure times down to the order of nanoseconds can be achieved. In such cases, unwanted movements become less important. The application of pulse lasers therefore substantially reduces the stringent stability requirements.

6.6 HOLOGRAPHIC INTERFEROMETRY

In Section 3.4 we mentioned an imaginary experiment where two waves reflected from two identical objects could interfere. With the method of holography now at hand, we are able to realize this type of experiment by storing the wavefront scattered from an object in a hologram. We then can recreate this wavefront by hologram reconstruction, where and when we choose. For instance, we can let it interfere with the wave scattered from the object in a deformed state. This technique belongs to the field of holographic interferometry (Horman 1965; Powell and Stetson 1965; Burch 1965;Haines and Hildebrand 1965; Vest 1979; Erf 1974; Jones and Wykes 1989). In the case of static deformations, the methods can be grouped into two procedures, double-exposure and real-time interferometry.

6.6.1 Double-exposure interferometry

In this method, two exposures of the object are made on the same hologram. This might be recordings before and after the object has been subject to load by, for instance, external forces or two other object states that are to be compared. By reconstructing the hologram, the two waves scattered from the object in its two states will be reconstructed simultaneously and interfere. This double-exposed hologram can be stored and later reconstructed for analysis of the registrated deformation at the time appropriate for the investigator. If a lot of different states of the object (e.g. different load levels) are to be investigated, many holograms have to be recorded, which makes the method time-consuming and elaborate.

6.6.2 Real-time Interferometry

In this method, a single recording of the object in its reference state is made. Then the hologram is processed and replaced in the same position as in the recording. By looking through the hologram we are now able to observe the interference between the reconstructed object wave and the wave from the real object in its original position. Thus we are able to follow the deformation as it develops in real time by observing the changes in the interference pattern. These changes might be recorded on film for later playback and analysis. A disadvantage of the method is that the hologram must be replaced in its original position with very high accuracy. This can be overcome by developing the hologram *in situ* in a transparent cuvette or using a thermoplastic film, see Section 5.4.2. Also the contrast of the interference fringes is not as good as in the double-exposure method.

6.6.3 Analysis of Interferograms

As we have seen, holographic interferometry enables the wave scattered from the object in its reference state described by the field amplitude $u_1 = U_1 e^{i\phi_1}$ and the wave scattered from the object in a deformed state described by the field amplitude $u_2 = U_2 e^{i\phi_2}$ to occur simultaneously. The actual deformations will be so small that we can put $U_1 = U_2 = U$. These two waves will form an interference pattern in the usual way given by

$$I = 2U^2[1 + \cos \Delta\phi] \qquad (6.17)$$

where

$$\Delta\phi = \phi_1 - \phi_2 \qquad (6.18)$$

The problem is then to find the relation between $\Delta\phi$ and the deformation.

Consider Figure 6.5 where a point O on the object is moved along the displacement vector **d** to the point O′ due to a deformation of the object. The object is illuminated by a plane wave (point source placed at infinity) which propagation direction $\mathbf{n}_1$ makes an angle θ_1 with the displacement vector **d**. Assume that we are looking through the hologram from infinity along the direction $\mathbf{n}_2$ making an angle θ_2 with **d**. We realize that the geometrical path length from the light source via the object point to the point of observation will be different before and after the deformation has taken place. In our case this difference is equal to the path length AO + OB which by applying simple trigonometry becomes equal to

$$d(\cos \theta_1 + \cos \theta_2) \qquad (6.19)$$

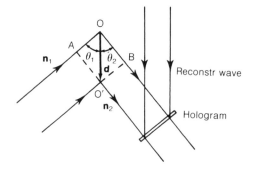

Fig. 6.5

From section 1.3 we know that the phase difference $\Delta\phi$ is equal to the path length multiplied by the wave number k:

$$\Delta\phi = kd(\cos\theta_1 + \cos\theta_2) \tag{6.20}$$

In Figure 6.6 a portion of Figure 6.5 is redrawn and the line bisecting the angle 2θ between $\mathbf{n}_1$ and $\mathbf{n}_2$ is introduced. This bisector is inclined at an angle γ to d. By using trigonometric formulas we find the geometry factor g to be

$$g = \cos\theta_1 + \cos\theta_2 = 2\cos\gamma\cos\theta \tag{6.21}$$

which yields

$$\Delta\phi = (2\pi/\lambda)2d(\cos\gamma)\cos\theta \tag{6.22}$$

By inserting Equation (6.22) into Equation (6.17) we find that the interference pattern has a maximum (bright fringe) whenever

$$\Delta\phi = (2\pi/\lambda)2d(\cos\gamma)\cos\theta = n\,2\pi \qquad \text{for} \quad n = 0, 1, 2, \ldots$$

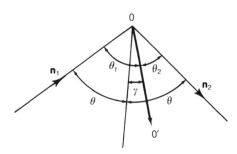

Fig. 6.6

i.e. when

$$d \cos \gamma = \frac{n\lambda}{2 \cos \theta} \qquad (6.23)$$

and a minimum (dark fringe) whenever

$$\Delta\phi = (2\pi/\lambda)2d(\cos \gamma) \cos \theta = (2n + 1)\pi \qquad \text{for} \quad n = 0, 1, 2, \ldots$$

i.e. when

$$d \cos \gamma = \frac{2n + 1}{2} \frac{\lambda}{2 \cos \theta} \qquad (6.24)$$

Here $d \cos \gamma$ is the component of the displacement vector onto the line bisecting the angle between the illumination and observation directons. This applies also when d do not lie in the plane defined by $\mathbf{n}_1$ and $\mathbf{n}_2$ as in Figure 6.6.

When interpreting interference patterns (also called interferograms) due to deformations of extended objects, we therefore can imagine the space to be filled with equispaced parallel planes which are normal to the bisector of $\mathbf{n}_1$ and $\mathbf{n}_2$ with a spacing equal to $\lambda/(2 \cos \theta)$. Each time the surface of the deformed object intersects one of these planes we get a bright (or dark) fringe. To measure the deformation at a given point, one therefore simply has to count the number of fringes and multiply it by $\lambda/(2 \cos \theta)$. This is illustrated in Figure 6.7 which shows the interference pattern seen on a beam that is rotated as a rigid body about an axis (the y-axis).

Highest sensitivity is obtained when both the illumination and observation directions are parallel to the displacement vector, i.e. when $\gamma = \theta_1 = \theta_2 = \theta = 0$. The displacement corresponding to one fringe spacing is then

$$d = \lambda/2$$

from Equation (6.24) we see that the first-order dark fringe ($n = 0$) then occurs at

$$d_{\min} = \lambda/4$$

Fig. 6.7

For the He–Ne laser wavelength $\lambda = 0.6328\,\mu$m, this corresponds to a displacement equal to $0.15\,\mu$m which therefore gives a representative figure for the sensitivity of standard holographic interferometry.

To obtain the best measurement accuracy when analysing holographic interferograms, the fringe positions should be determined with the highest possible accuracy (or fractional fringe width). This can be achieved in many different ways as will be discussed in chapter 11. Here we mention one method based on the heterodyne principle (see section 3.5.4) called heterodyne holographic interferometry (HHI) (Dandiker and Thalmann 1985). In this method the two wavefronts corresponding to the object in its two states have different frequencies v_1 and v_2. Equation (6.17) then becomes

$$I = 2U^2\{1 + \cos[2\pi(v_1 - v_2)t + \Delta\phi]\} \tag{6.25}$$

where $v_1 - v_2$ corresponds to the intermediate frequency of Equation (3.38) and the analysis to recover $\Delta\phi$ follows the principle described in Section 3.5.4. Such a measurement can be made to an accuracy of typically 10^{-1} radians or $3\cdot10^{-3}$ of a cycle and hence enables very small deformations to be resolved. Moreover the measurement is independent of U which affects only the amplitude of the signal. Heterodyning may be used in both real-time and double exposure holography. In real-time HHI, the 'live' object beam and the reconstructing beam have different frequencies, whilst in double exposure HHI, a single frequency is used to make both exposures but two separate reference beams are employed. At the reconstruction stage these two reference beams are set at different frequencies. In both methods two detectors are required which are used in one of two ways. For the first of these, one of the detectors is tracked across the fringe pattern whilst the other is held static and hence generates the fixed reference frequency. Alternatively the two detectors may be maintained at a fixed distance with respect to one another and both tracked across the fringe field. This enables a differential measurement to be made and hence the variations in the gradient of $\Delta\phi$ is found. In practice the two frequencies are obtained by splitting a single continuous laser beam and passing the two beams through separate optoacoustical modulators.

When the point source and the point of observation are placed at finite distances from the object, the illumination and observation directions ($\mathbf{n}_1$ and $\mathbf{n}_2$) will vary across the object surface. This will turn the above-mentioned plane equispaced surfaces into equidistant ellipsoids with the point source and the point of observation as foci. The sensitivity (displacement per fringe) then varies across the object surface but for small objects and reasonably long distances to the source point and the observation point this variation will be quite small. As can be seen from Equation (6.22) and Figure 6.7 the method is insensitive to displacements parallel to the equispaced planes or along the ellipsoids in the general case.

This should be kept in mind when making a set-up for holographic interferometry. The concept called the holodiagram (which is essentially the ellipsoids mentioned above) developed by Abramson (1970, 1972) could be helpful in this respect.

Figure 6.8 shows some typical examples of interferograms obtained by means of holographic interferometry on different objects. It should be noted that holographic interferometry is incapable of measuring surface contour deviations between two different objects as pointed out in Section 3.4. It is therefore also impossible to measure deformations if the microstructure of the object changes drastically, as for example in plastic deformations.

Fig. 6.8.   *(a)*

Fig. 6.8.   *(b)*

Fig. 6.8. *Examples of holographic interferograms. (a) Deflection of a rectangular plate fastened with five struts and subjected to a uniform pressure. From Wilson et al. 1971. (Photograph courtesy of Dr A. D. Wilson, Thomas J. Watson Research Center, Yorktown Heights, New York. Reproduced by permission of SEM.) (b) Detection of debonded region of a honeycomb construction panel. From Vest 1979. (Reproduced by permission of John Wiley & Sons Inc.) (c) A bullet in flight observed through a doubly-exposed hologram. From Collier et al. 1971. (Reproduced by permission of Dr R. F. Wuerker, TRW Inc.) (d) Holographic reconstruction of a solid turbine blade illustrating (A) the first flexural resonance at 981 Hz, (B) a second-order flexural resonance at 4624 hz, and the 2nd and 3rd torsional resonances at (C) 6406 Hz. From Erf 1974. (Photograph courtesy at Dr R. K. Erf, United Technologies Research Center, East Hartiord, Connecticut. Reproduced by permission of Academic Press, New York)*

(d)

Fig. 6.8. *(d)*

6.6.4 Localization of interference fringes

Another difficulty in evaluating holographic interferograms comes from the phenomenon of fringe localization. It is an annoying fact that the fringes in holographic interferometry only in special cases are localized on the object surface.

This means that when imaging an interferogram by focusing on the object, the fringes may be completely lost because they focus (localize) in a plane (which might be curved) which lies far away from the object surface. That the fringes are localized in a certain plane means that they have maximum contrast or visibility in that plane. Loss of friges in the plane of the object or other planes therefore means that their contrast is too low to be detected in that plane.

To see this, we must remember that the interferogram is formed by interference between light scattered from pairs of corresponding points on the object surface in the first and second exposure. Interference between non-identical points on the two displaced surfaces will contribute a random noisy background to the interferogram, thereby reducing its contrast.

In the analysis below, we shall show that this problem might be overcome by stopping down the aperture of the imaging system, thereby increasing the depth of focus to obtain simultaneously an image of sufficient quality of both the fringe pattern and the object surface. For more details, see Vest (1979), Walles (1969) and Molin and Stetson (1971).

Consider Figure 6.9 where the holographic interferogram due to displacement of the object surface (drawn as a single surface in the figure) is imaged onto a viewing screen by a lens through a rectangular aperture. Assume that the lens images a plane containing the

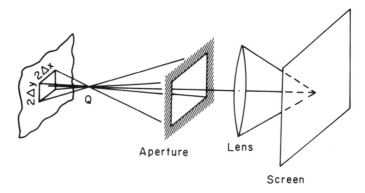

Aperture   Lens

Screen

Fig. 6.9

point Q onto the viewing screen. The central ray passing through Q emanates from a point $P(x_0, y_0)$ on the object surface. The intensity at Q is the integral of the intensity of all the ray pairs within the ray cone defined by the aperture. This cone subtends a rectangular area of dimension $2\Delta x$, $2\Delta y$ on the object surface, hence the intensity at Q is

$$I(Q) = \int_{x_0 - \Delta x}^{x_0 + \Delta x} \int_{y_0 - \Delta y}^{y_0 + \Delta y} [1 + \cos \delta(x, y)] \, dx \, dy \tag{6.26}$$

where

$$\delta(x, y) = kd(\cos \theta_1 + \cos \theta_2) \tag{6.27}$$

is the phase difference due to the displacement of each point on the object between the two exposures (see Equation (6.20)). For small Δx, Δy, $\delta(x, y)$ can be approximated by the initial terms of a Taylor series expansion about x_0, y_0:

$$\delta(x, y) = \delta(x_0, y_0) + (x - x_0)\frac{\partial \delta}{\partial x}\bigg|_{x_0, y_0} + (y - y_0)\frac{\partial \delta}{\partial y}\bigg|_{x_0, y_0}$$

$$= \delta_0 + (x - x_0)\delta_0^x + (y - y_0)\delta_0^y \tag{6.28}$$

where δ_0^x and δ_0^y denote partial derivatives of δ at x_0, y_0 with respect to x and y respectively.

Substituting Equation (6.28) into Equation (6.26) and evaluating the integral yields

$$I(Q) = 4\Delta x\Delta y\left[1 + \frac{\sin(\delta_0^x\Delta x)}{\delta_0^x\Delta x}\frac{\sin(\delta_0^y\Delta y)}{\delta_0^y\Delta y}\cos \delta_0\right] \tag{6.29}$$

The contrast of this intensity distribution is found from the definition of contrast or visibility (cf. Equation (3.8)):

$$V = \frac{I_{max} - I_{min}}{I_{max} + I_{min}} \tag{6.30}$$

which by putting $\cos \delta_0 = 1$ for I_{max} and $\cos \delta_0 = -1$ for I_{min} in Equation (6.29) gives

$$V = \left|\frac{\sin(\delta_0^x\Delta x)}{\delta_0^x\Delta x}\frac{\sin(\delta_0^y\Delta y)}{\delta_0^y\Delta y}\right| \tag{6.31}$$

From this expression we see that the visibility is equal to 1 for δ_0^x or $\delta_0^y = 0$, i.e. when the variations in phase over the viewing cone are minimized. This defines the area of fringe localization. If the value of $\delta_0^x \Delta x$ or $\delta_0^y \Delta y$ increases rapidly as the distance z is changed from the localization distance, the region of fringe localization is sharply defined. If $\delta_0^x \Delta x$ or $\delta_0^y \Delta y$ increases slowly as z is varied, the region of localization is broad. Since δ_0^x and δ_0^y are determined by the system geometry and object displacement field, we see from Equation (6.31) that the sharpness of localization can be controlled by the viewing aperture $\Delta x \Delta y$. Therefore, by decreasing the aperture, relatively distinct fringes can be observed over an extended region and fringes can be seen in the plane of the object even though it is at some distance from the region of localization.

We also see from Equation (6.31) that the contrast is a periodic function of $\delta_0^x \Delta x$, $\delta_0^y \Delta y$. As we move away from the region of localization, the contrast therefore can assume periodic maxima but with a much lower value due to the sinc-function dependence (see Figure B.1.c).

From Equation (6.27) we see that δ is dependent on the illumination and observation directions as well as the displacement vector d. The determination of the regions of fringe localization must therefore be calculated in each separate case by maximizing Equation (6.31). Here we shall not go into such detailed calculations but merely quote two general results which are:

(1) For a rigid body translation the fringes localize at infinity.
(2) For a rigid body rotation about an axis inside the object's surface and normal to the illumination and observation directions the fringes localize on the object.

In conclusion, fringe localization rarely poses any problem to the experimentalist since it can be solved by decreasing the viewing aperture. On the other hand, the phenomenon can be taken advantage of since it conveys additional information about the deformation under investigation.

6.7 HOLOGRAPHIC VIBRATION ANALYSIS

Up to now we have been dealing with holographic interferometry applied on static deformations. In this section we shall show that this method can also be applied to vibrating objects.

Assume that the object point in Figure 6.5 executes harmonic vibrations given by

$$d(x, t) = D(x) \cos \omega t \tag{6.32}$$

where $D(x)$ is the amplitude, x represents the spatial coordinates of the object point and ω is the vibration frequency. Light scattered from this point can be described by a field amplitude in the hologram plane given by

$$u_0(x, t) = U_0(x) e^{i\phi} \tag{6.33}$$

where

$$\phi = kgd(x, t). \tag{6.34}$$

Here $g = \cos \theta_1 + \cos \theta_2$ is the geometry factor determined by the illumination and observation directions in the same way as in the static case.

Let u_o of Equation (6.33) be the object wave in a hologram recording and u the reference wave. Just as in the static case, the reconstructed wave will be given by the last term of Equation (6.3). However, since u_o is time-varying during the exposure, the recorded intensity distribution (Equation (6.1)) and hence the hologram amplitude transmittance (Equation (6.2)) will be averaged over the exposure time. The reconstructed object wave therefore becomes equal to

$$u_a = \alpha|u|^2\bar{u}_o \tag{6.35}$$

where the bar denotes time averaging. For exposure times much longer than the vibrating period of the object (Løkberg 1979) this time averaging is equivalent to averaging over one vibration period T. Therefore

$$u_a = \alpha|u|^2 \frac{1}{T} \int_0^T u_o(x,t)\,\mathrm{d}t$$

$$= \alpha|u|^2 U_o \frac{1}{T} \int_0^T e^{ikgD(x)\cos\omega t}\,\mathrm{d}t = \alpha|u|^2 U_o J_0[kgD(x)] \tag{6.36}$$

where J_0 is the zeroth-order Bessel function. Here we have applied the relation

$$\frac{1}{2\pi} \int_0^{2\pi} e^{i\eta\cos\xi}\,\mathrm{d}\xi = J_0(\eta) \tag{6.37}$$

The observable intensity distribution of the reconstructed wave becomes

$$I_a = |u_a|^2 = KJ_0^2[kgD(x)] \tag{6.38}$$

where all constants are gathered into a common constant K.

As an illustrative example, consider Figure 6.10(a) where a bar is vibrating as a rigid body about an axis. In Figure 6.10(b) we have drawn the Bessel function squared which represents the intensity distribution that will be observed in the reconstruction of a hologram recording of the vibrating bar. We see that the region around the axis which is at rest (the nodal point) will show up as a bright zero-order fringe of much higher intensity

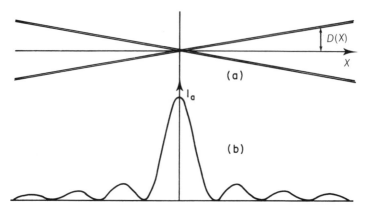

Fig. 6.10. *(a) Vibration of a bar about an axis. (b) Intensity distribution of the resulting time-average holographic recording*

than the higher-order bright fringes farther along the bar. This is in contrast to the cos-fringes obtained in the case of a static deformation where all bright fringes have equal intensity.

To find the vibration amplitude at the higher-order fringes we consult a table of Bessel function values from which we find that

$$J_0^2(\eta) = \text{max for } \eta = 0, 3.83, 7.02, 10.17, 13.32, 16.47, \ldots$$
$$J_0^2(\eta) = 0 \text{ for } \eta = 2.40, 5.52, 8.65, 11.79, 14.93, \ldots$$

For $g = 2$ (illumination and observation directions parallel to the displacement) and $\lambda = 632.8$ nm (wavelength of the He–Ne laser) this gives:

Bright fringes when $D(x) = 0, 0.19, 0.35, 0.51, 0.67, 0.83, \ldots$ [µm]

Dark fringes when $D(x) = 0.12, 0.28, 0.44, 0.59, 0.75, \ldots$ [µm]

By means of standard holographic techniques one has been able to observe fringes up to the 50th order by this method. A very detailed map of the amplitude distribution is therefore obtained. The frequency range of holographic vibration measurements is only limited by the method of object excitation. Using piezoelectric transducers for excitation, values of hundreds of kilohertz are easily obtained.

The method described above is called the time-average method. Another method similar to the real-time method for static displacements can also be applied. It consists of first recording a hologram while the object is at rest, then replacing the hologram in its original position and observing the resulting fringe pattern when the object vibrates. The contrast of this pattern is very low due to the resulting $(1 - J_0)$-dependence (Vest 1979). While observing this real-time pattern the laser beam can be chopped with the same frequency. This is equivalent to the real-time method for static displacements. Here one observes fringes due to the displacement of the object, when at rest and when illuminated by the light pulse. In that way, by slowly varying the phase between the light pulse and the object vibration, one can observe the vibration of the object in slow motion. Thus it is also possible to observe the phase of the displacements on the different parts of the object surface.

Another method is to phase modulate the reference wave (Aleksoff 1971, 1974). This can be done by placing a vibration mirror in the light path of the reference wave. The argument of the Bessel function of Equation (6.38) then changes from $(4\pi/\lambda)D(x)$ (for $g = 2$) to

$$\frac{4\pi}{\lambda}[D_o^2(x) + R^2 - 2D_o(x)R\cos(\psi_o(x) - \psi_r)]^{1/2} \tag{6.39}$$

where

$D_o(x)$ = the vibrating amplitude of the object,
R = the vibrating amplitude of the reference mirror,
$\psi_o(x) - \psi_r$ = the phase difference between the vibration of the object and reference mirror.

The result of this reference wave modulation is that the centre of the Bessel function is moved from the nodal points of the object to points that vibrate at the same amplitude and phase as the modulating mirror. These points then show up as zero-order bright fringe areas. In that way it is possible to extend the measurable amplitude range considerably, in practice up to about 10 µm. By varying the phase of the reference mirror it is also possible to trace out the areas of the object vibrating in the same phase as the reference mirror, thereby mapping the phase distribution of the object.

Reference mirror modulation can also be used to measure very small vibration amplitudes by moving the steepest part of the central maximum of the Bessel function to coincide with zero object vibration amplitude (Metherell *et al.* 1969; Hogmoen and Løkberg 1976).

Application of a TV camera as the recording medium gives a very versatile instrument for vibration analysis using the methods described above. This will be treated in more detail in Section 6.8.4.

6.8 SPECIAL TECHNIQUES IN HOLOGRAPHIC INTERFEROMETRY

6.8.1 Introduction

As mentioned in Section 6.5, holography is very sensitive to unwanted disturbances. When applying holographic interferometry to strained solid objects, the main aim is to measure the resulting local deformations on the object surface. However, depending on the loading conditions, the object will very often also undergo rigid body movements and these are commonly of much greater magnitude than the local deformations.

Such rigid body displacements may therefore often be too great to be measured by holographic interferometry or they will completely mask the local deformations searched for in the resulting interference fringe pattern. In practical applications, methods for separating, isolating or compensating for such unwanted rigid body motions are therefore badly needed. In the following subsections we shall describe some techniques which have proved useful in this respect.

6.8.2 Sandwich holography

In this method, originally devised by Abramson (1974, 1981), two separate holograms are put together in a sandwich (see Figure 6.11). The object in its reference position is recorded on the back hologram whereafter the deformed object is recorded on the front hologram. After development, these two holograms are reconstructed simultaneously.

If the object has undergone an unwanted tilt between the two exposures, this can be compensated for by tilting the sandwich holograms leaving an interferogram due only to the deformation of the object.

To see how this works, consider Figure 6.11(a) where a light ray from an object point is incident on the two holograms at an angle ψ_1. The reference wave is a plane wave incident at an angle θ_1, and the two holograms are separated by a distance s. Now, suppose that the object is unaltered between the exposures on the back and front holograms. In reconstruction, the path-length difference (measured from the reconstruction wave) between the rays reconstructed from the back and the front holograms is seen to be

$$AD - C_1D = r - r\cos(\psi_1 + \theta_1) \tag{6.40}$$

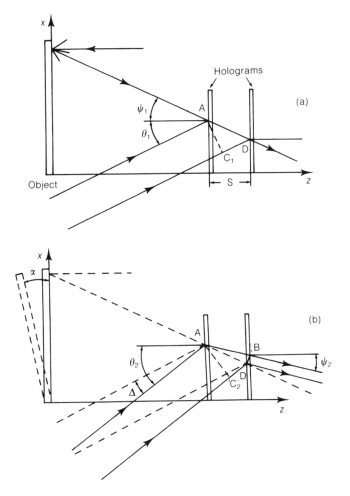

Fig. 6.11

where

$$r = s/\cos\psi_1 \qquad (6.41)$$

Since the object is unaltered between the two recordings, the two reconstructed waves must be in phase. Equation (6.40) thus represents zero phase for the two reconstructed waves. It should be noted here that when recording the back hologram a 'dummy' hologram is placed in the position of the front hologram to have the same phase relations as in the reconstruction.

Tilting of the sandwich hologram is equivalent to changing the angle of incidence of the reconstruction wave. This situation is sketched in Figure 6.11(b) where the angle of incidence of the reconstruction wave is increased to $\theta_2 = \theta_1 + \Delta$. The path-length difference between the rays reconstructed from the back and front holograms now is seen to be

$$AB - C_2D = r\cos(\psi_1 - \psi_2) - r\cos(\psi_1 + \theta_2) \qquad (6.42)$$

where ψ_2 is the angle of emergence of the reconstructed waves. The net path length difference between the reconstructed waves introduced by tilting the sandwich is therefore found by taking the difference of Equation (6.40) and (6.42) which gives

$$l = AB - C_2D - (AD - C_1D) \tag{6.43}$$

$$= \frac{s}{\cos\psi_1}[\cos(\psi_1 - \psi_2) - \cos(\psi_1 + \theta_2) - 1 + \cos(\psi_1 + \theta_1)]$$

In recording, the object and reference rays will locally form a regular sinusoidal grating. The period of this grating will determine the direction of the recostructed ray according to the grating equation (Equation (4.19)). The angles involved must therefore obey the relation

$$\sin\psi_1 + \sin\theta_1 = \sin\psi_2 + \sin\theta_2 \tag{6.44}$$

Combination of Equation (6.43) and (6.44) gives, for the path-length difference introduced by tilting the sandwich holograms,

$$l = s\left[(2\sin\theta_2 - 2\sin\theta_1 - \sin\psi_1)\tan\psi_1 + \cos\psi_2 - \frac{1}{\cos\psi_1} + (\cos\theta_1 - \cos\theta_2)\right] \tag{6.45}$$

An important point to note is that this expression depends almost linearly on $\tan\psi_1$, because tilting of the object will introduce a phase variation proportional to $\tan\psi_1$, the angle of observation. For small observation angles, Equation (6.45) becomes

$$l = s[2(\sin\theta_2 - \sin\theta_1)\psi_1 + (\cos\theta_1 - \cos\theta_2)] \tag{6.46}$$

and we have complete proportionality. For small rotations Δ of the hologram we further have

$$l = s\Delta(2\psi_1\cos\theta_1 + \sin\theta_1) \tag{6.47}$$

which for $\theta_1 = 0$ gives

$$l = 2s\Delta\psi_1 \tag{6.48}$$

An object in the xy-plane tilted an angle α about the y-axis will introduce a path-length change equal to $2\alpha x$ where x is the point of observation.

To cancel this tilt by rotating the sandwich we therefore must have (for $\theta_1 = 0$)

$$2\alpha x = l = 2s\Delta\psi_1 = 2s\Delta\frac{x}{z} \tag{6.49}$$

where z is the distance from the object to the observer. The sandwich thus must be rotated an angle

$$\Delta = \frac{z}{s}\alpha \tag{6.50}$$

Figure 6.12 shows a striking application of sandwich holography. It is a two-metre high milling machine where a force between the tool and the workpiece is applied. Figure 6.12(a) shows the reconstruction without tilting of the sandwich. In Figure 6.12(b) the

Fig. 6.12. *(a) Reconstruction of a sandwich hologram recording of a milling machine where a force between the tool and the workpiece has been aplied between the two exposures. (b) Same as (a) after tilting of the sandwich hologram. From Abramson 1981. (Reproduced by permission of Professor N. Abramson, Royal Institute of Technology, Stockholm)*

sandwich hologram is rotated until a minimum number of fringes appear on the machine head and the sole deformation of the head is displayed.

In the lower right corner of the picture is placed a reference surface on which the number of fringes indicate the rotation of the hologram.

6.8.3 Reflecton holography

In this method, rigid-body movements are eliminated by clamping the hologram plate (magnetically or mechanically) to the object under study, as shown in Figure 6.13(a). The hologram, with its emulsion facing the object, is illuminated by a plane wave at normal incidence. In its first pass through the emulsion, this wave acts as the reference wave. Then

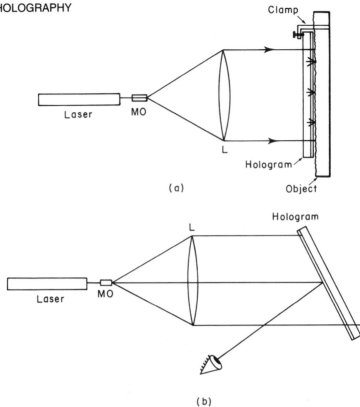

Fig. 6.13. *(a) Recording and (b) reconstruction of a reflection hologram in contact with the test specimen. MO = microscope objective*

it is reflected back from the object surface and acts as the object wave in its second pass. In this way a reflection or volume hologram is obtained (see Section 6.4). To achieve this, a thick emulsion must be applied. The Agfa-Gevaert Holotest 8E75 HD-plates have proved useful for this purpose.

As mentioned in Section 6.4, a reflection hologram can be reconstructed in white light. When using a laser for reconstruction, as in Figure 6.13(b), one must take account of the shrinkage of the emulsion. This means that when using normal photographic processing, a laser of shorter wavelength than in the recording must be applied. By re-swelling the emulsion after processing, the hologram can be reconstructed with the same wavelength as in the recording.

When using a collimated plane wave in the recording, the resulting interferogram becomes insensitive to rigid-body translations parallel and normal to the illumination direction. Also in-plane rotation has no influence on the result.

Rigid-body rotation around an axis perpendicular to the illumination direction, however, will produce additional fringes. This can be seen from Figure 6.14 where a point P on the object is rotated (together with the hologram) an angle α to point P' around an axis R which is normal to the plane of the paper. Assume that we observe the interferogram at an angle β to the illumination direction. In the first exposure the path length from a wavefront in the hologram plane and back to the same wavefront is IP + PH. After

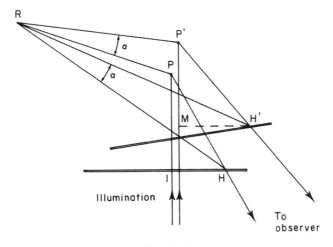

Fig. 6.14

rotation, the point H on the hologram is moved to H'. The corresponding path length is now measured from the wavefront intersecting H', which gives $MP' + P'H'$, where $P'H' = PH$ and

$$MP' = P'H' \cos(\alpha + \beta) = PH \cos(\alpha + \beta)$$

The path difference due to the rotation therefore becomes

$$l = IP + PH - MP' - P'H'$$

$$= PH[\cos\beta - \cos(\alpha + \beta)] = z_p\left[1 - \frac{\cos(\alpha + \beta)}{\cos\beta}\right] \tag{6.51}$$

where $z_p = IP$ is the distance between P and the hologram along the illumination direction. For small α we get

$$l \simeq z_p\left(1 - \frac{\cos\beta - \alpha\sin\beta}{\cos\beta}\right) = z_p\alpha\tan\beta \tag{6.52}$$

From this expression we see that $l = 0$, for $\alpha \neq 0$ (i.e. no additional fringes caused by rotation) when

$z_p = 0$, i.e. object in contact with hologram,
$\beta = 0$, i.e. the observation direction coincides with the illumination direction.

The method of clamping the hologram to the object can also be applied to transparent objects. The set-up is then as shown Figure 6.15. We then no longer have a reflection hologram. the surface of the specimen should have some roughness to scatter a portion of the light diffusely.

When exposing the film emulsion to the wave scattered from the object we also have recorded the speckle pattern of the object in the same way as in speckle photography, see Chapter 8. The double-exposed hologram can therefore be used to form the so-called Young fringes (see Section 8.3.2) or Fourier fringes (see Section 8.3.1) by means of the

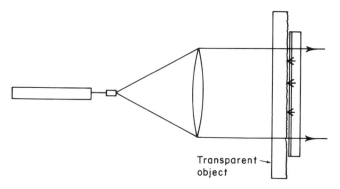

Fig. 6.15

set-up shown in Figure 4.18. The resulting fringe pattern gives the in-plane displacements of the object surface. Such a contact hologram can therefore give the three Cartesian components of any displacement vector. In contrast to ordinary speckle photography, the film will get an additional uniform blackening due to the reference wave. The contrast of the speckle pattern might therefore be very low, making it difficult or even impossible to extract the in-plane displacement fringes. It might therefore be necessary to make a separate recording for speckle correlation purposes and to remove the unwanted uniform blackening by a reducer step of the Farmer type in the processing of the film plate.

Figure 6.16 shows an example of the result obtained by this method reproduced from the pioneering work of Boone (1974, 1975). The object is a short, notched PMMA beam loaded by bending. Figure 6.16(a) shows the holographic out-of-plane displacement fringes while Figure 6.16(b) and (c) show the in plane Fourier fringes. The holograms were recorded using a He–Ne laser and reconstructed by a wavelength of 590 nm from a dye laser. The speckle patterns were recorded at a wavelength of 514.5 nm using an argon laser.

As we have seen, the application of this technique is quite simple, the number of optical components being kept to a minimum. It has been successfully applied to test specimens under tension in rigs for fatigue tests on the factory floor. A disadvantage is that it can be applied only to relatively flat objects and the measuring area is limited by the aperture of the collimating lens and/or the area of the hologram plate.

6.8.4   TV holography (ESPI)

In this technique (also called electronic speckle pattern interferometry, ESPI) the holographic film is replaced by a TV camera as the recording medium (Jones and Wykes 1989). Obviously, the target of a TV camera can be used neither as a holographic storage medium nor for optical reconstruction of a hologram. Therefore the reconstruction process is performed electronically and the object is imaged onto the TV target. Because of the rather low resolution of a standard TV target (about 30 lines/mm), the angle between the object and reference waves has to be as small as possible. This means that the reference wave is made in-line with the object wave. A typical TV holography set-up therefore looks like that given in Figure 6.17. Here a reference wave modulating mirror (M_1) and a chopper

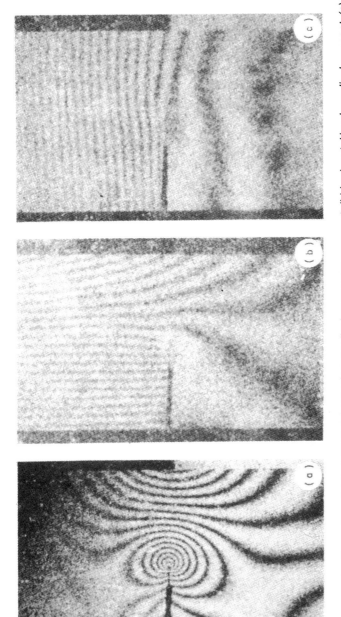

Fig. 6.16. *Displacement patterns for notch bend test: (a) out-of-plane displacement; (b) horizontal in-plane displacement; (c) vertical in-plane displacement. (Photograph courtesy by Professor. Boone, laboratorium Soete voor werstand van materialen en lastechniek, University of Ghent. From Boone 1975. Reproduced by permission of Taylor & Francis Ltd)*

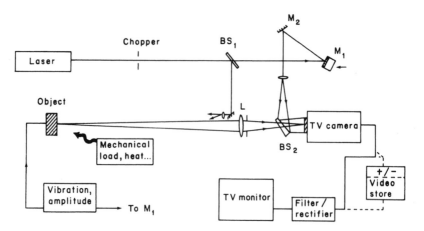

Fig. 6.17. *TV-holography set-up. From Lokberg 1980. (Reproduced by permission of Prof. O. J. Løkberg, Norwegian Institute of Technology, Trondheim)*

are included, which are necessary only for specal purposes in vibration analysis (see Section 6.7).

The basic principles of ESPI were developed almost simultaneously by Macovski et al (1971) in the USA, Schwomma (1972) in Austria and Butters and Leendertz (1971) in England. Later the group of Løkberg (1980) in Norway contributed significantly to the field, especially in vibration analysis (Løkberg and Slettemoen 1987).

When the system in Figure 6.17 is applied to vibration analysis the video store is not needed. As in the analysis in Section 6.7, assume that the object and reference waves on the TV target are described by

$$u_o = U_o e^{i\phi_o} \tag{6.53a}$$

and

$$u = U e^{i\phi} \tag{6.53b}$$

respectively. For a harmonically vibrating object we have (see equation (6.34) for $g = 2$)

$$\phi_0 = 2kD(x) \cos \omega t \tag{6.54}$$

where $D(x)$ is the vibration amplitude at the point of spatial coordinates x and ω is the vibration frequency. The intensity distribution over the TV-target becomes

$$I(x,t) = U^2 + U_o^2 + 2UU_o \cos[\phi - 2kD(x) \cos \omega t] \tag{6.55}$$

This spatial intensity distribution is converted into a corresponding time-varying video signal. When the vibration frequency is much higher than the frame frequency of the TV system ($\frac{1}{25}$ s, European standard), the intensity observed on the monitor is proportional to Equation (6.55) averaged over one vibration period, i.e. (cf. Equation (6.36))

$$I = \overline{U^2} + \overline{U_o^2} + \overline{2UU_o \cos \phi} J_0(2kD(x)) \tag{6.56}$$

where J_0 is the zeroth-order Bessel function and the bars denote time average. Before being displayed on the monitor, the video signal is high-pass filtered and rectified. In the filtering process, the first two terms of Equation (6.56) are removed. After full-wave rectifying we thus are left with

$$I = 2|\overline{UU_o \cos\phi} \, J_0(2kD(x))| \tag{6.57}$$

Actually, ϕ represents the phase difference between the reference wave and the wave scattered from the object in its stationary state. The term $UU_o \cos\phi$ therefore represents a speckle pattern and the J_0-function is said to modulate this speckle pattern. Equation (6.57) is quite analogous to Equation (6.38) except that we get a $|J_0|$-dependence instead of a J_0^2-dependence. The maxima and zeros of the intensity distributions have, however, the same locations in the two cases. A time-average recording of a vibrating turbine blade therefore looks like that shown in Figure 6,18(a) when applying ordinary holography, and that in Figure 6.18(b) when applying TV holography. We see that the main difference in the two fringe patterns is the speckled appearance of the TV holography picture.

When applied to static deformations, the video store in Figure 6.17 must be included. This could be a video tape or disc, or most commonly, a frame grabber (see Section 10.1) in which case the video signal is digitized by an analogue-to-digital converter. Assume that the wave scattered from the object in its initial state at a point on the TV target is described by

$$u_1 = U_o e^{i\phi_o} \tag{6.58}$$

After deformation, this wave is changed to

$$u_2 = U_o e^{i(\phi_o + 2kd)} \tag{6.59}$$

where d is the out of plane displacement and where we have assumed equal field amplitudes in the two cases. Before deformation, the intensity distribution on the TV target is given by

$$I_1 = U^2 + U_o^2 + 2UU_o \cos(\phi - \phi_o) \tag{6.60}$$

Holography   ESPI

(a)   (b)

Fig. 6.18. *(a) Ordinary holographic and (b) TV-holographic recording of a vibrating turbine blade. (Reproduced by permission of Prof. O. J. Løkberg, Norwegian Institute of Technology, Trondheim)* where U and ϕ are the amplitude and phase of the reference wave. This distribution is converted into a corresponding video signal and stored in the memory. After the deformation, the intensity and corresponding video signal is given by

$$I_2 = U^2 + U_o^2 + 2UU_o\cos(\phi - \phi_0 - 2kd) \tag{6.61}$$

These two signals are then subtracted in real time and rectified, resulting in an intensity distribution on the monitor proportional to

$$\begin{aligned} I_1 - I_2 &= 2|UU_0|[\cos(\phi - \phi_0) - \cos(\phi - \phi_0 - 2kd)]| \\ &= 4|UU_0|\sin(\phi - \phi_0 - kd)\sin(kd)| \end{aligned} \tag{6.62}$$

The difference signal is also high-pass filtered, removing any unwanted background signal due to slow spatial variations in the reference wave. Apart from the speckle pattern due to the random phase fluctuations $\phi - \phi_0$ between the object and reference fields, this gives the same fringe patters as when using ordinary holography to static deformations. The dark and bright fringes are, however, interchanged, for example the zero-order dark fringe corresponds to zero displacement.

This TV holography system has a lot of advantages. In the first place, the cumbersome, time-consuming development process of the hologram is omitted. The exposure time is quite short ($\frac{1}{25}$ s), relaxing the stability requirements, and one gets a new hologram each $\frac{1}{25}$ s. Among other things, this means that an unsuccessful recording does not have the same serious consequences and the set-up can be optimized very quickly. A lot of loading conditions can be examined during a relatively short time period. Time-average recordings of vibrating objects at different excitation levels and different frequencies are easily performed.

The interferograms can be photographed directly from the monitor screen or recorded on video tape for later analysis and documentation. TV holography is extremely useful for applications of the reference wave modulation and stroboscopic holography techniques mentioned in Section 6.7. In this way, vibration amplitudes down to a couple of angstroms have been measured. The method has been applied to a lot of different objects varying from the human ear drum (Løkberg et al. 1979) to car bodies (Malmo and Vikhagen 1988).

When analysing static deformations, the real-time feature of TV holography makes it possible to compensate for rigid-body movements by tilting morrors in the illumination beam path until a minimum number of fringes appear on the monitor.

6.9 HOLOGRAPHIC INTERFEROMETRY OF TRANSPARENT OBJECTS

Up to now we have mainly considered the application of holographic interferometry to opaque objects. The method can, however, equally well be used for the analysis of transparent objects (Vest 1979). In fact, the set-up becomes slightly simpler, for example that shown in Figure 6.19.

The quantity actually measured by this method is the change in refractive index due to some change in the object volume. For three-dimensional objects, the corresponding phase difference is an integrated value through the object volume along the ray path. In

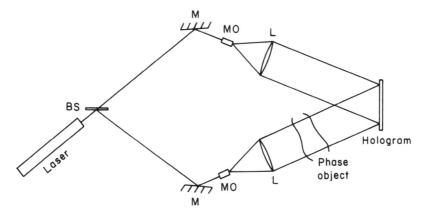

Fig. 6.19. *Holography set-up for transparent phase objects*

the same way as for opaque objects, the resulting interferogram is given as (cf. Equation (6.17))

$$I = 2U^2(1 + \cos \Delta\phi) \tag{6.63}$$

where $\Delta\phi$ is the phase difference in the two recordings. In most applications the refractive index during one exposure, say the first, is uniform and can be denoted by n_0. Then the difference in the general case is given as

$$\Delta\phi = k \int [n(x, y, z) - n_0]dz \tag{6.64}$$

where $n(x, y, z)$ is the refractive index distribution during the second exposure and where we have assumed plane wave object illumination along the z-axis. From this equation we see that it is impossible to determine the phase contribution from a single specific point inside the object volume. There are, however, special cases, for example refractive index variations only in the x- and y-directions, radially symmetric distributions, etc. for which Equation (6.64) can be explicitly solved.

Holographic interferometry of transparent phase objects has nevertheless become a versatile tool in research areas such as aerodynamics, heat transfer, plasma diagnostics and stress analysis of transparent models. The latter will be treated in more detail in Section 9.7.

Figure 6.8(c) shows a typical example from the field of aerodynamics. it is a two-exposure holographic interferogram of high speed air flow past a cone. The change in refractive index due to air compression can be found by counting the number of fringes starting from the tip of the cone.

We shall not here go into details about the various techniques used in this field. In fact, a lot of different methods have been applied. For reference, we merely quote some relations between the refractive index and some physical quantities appropriate to the different measurement problems.

In aerodynamics and flow visualization, interferometry is used to determine the distribution of density (the mass per unit volume) in a gas. Density, denoted by ρ, is related to the refractive index of the gas by the Gladstone–Dale equation:

$$n - 1 = K\rho \tag{6.65}$$

where K, the Gladstone-Dale constant, is a property of the gas. K is nearly independent of the wavelength of light and of temperature and pressure under moderate physical conditions.

In heat and mass transfer, interferometry is used to determine the spatial distribution of temperature or concentration of chemical species. For gases, the relation between temperature and refractive index is found by combining the ideal gas equation of state

$$\rho = \frac{MP}{RT} \tag{6.66}$$

and the Gladstone-Dale equation (Equation (6.65)) to yield

$$n - 1 = \frac{KMP}{RT} \tag{6.67}$$

Where P is the pressure, M is the molecular weight of the gas, T is the absolute temperature and R is the universal gas constant. The slope of a curve of refractive index versus temperature is therefore

$$\frac{\mathrm{d}n}{\mathrm{d}T} = -\frac{KMP}{RT^2} \tag{6.68}$$

For small temperature changes, the right-hand side of Equation (6.68) is approximately constant, giving a linear relation for the rate of change of refractive index with temperature. For liquids, the density is not a simple function of temperature so empirical relations must be found between n and T.

In plasma diagnostics, interferometry is used to determine the spatial distribution of densities of the plasma which is a collection of atoms, ions and electrons created by very high temperatures. The refractive index of a plasma is the sum of the refractive indices of these particles weighted by their number densities. For atoms and ions, the refractive index is given by the Gladstone-Dale relation, Equation (6.65). For electrons, the refractive index n_e is given as

$$n_e = \left(1 - \frac{N_e e^2 \lambda^2}{2\pi m_e c^2}\right)^{1/2} \tag{6.69}$$

where e is the charge and m_e the mass of an electron, c is the speed of light and N_e is the numer density of electrons. We see that this relation is strongly dependent on the wavelength λ.

PROBLEMS

6.1 In Section 3.3 we found an expression for the interfringe distance d_x measured along the x- axis when two plane waves are incident on the xy-plane, see Figure 3.2 and Equation (3.23).
Calculate the interfringe distance along the x-axis when the two plane waves are refracted into a medium (a hologram plate) of refractive index n.

6.2 Imagine that you cut a small piece out of a recorded hologram plate (you might have dropped the hologram glass plate on the floor!) and use this piece in the reconstruction. What do you see? Explain.

6.3 In Section 6.3 we stated that maximum diffraction efficiency for a thin amplitude hologram is obtained when the visibility (or contrast) of the intensity distribution is unity, which means that the ratio $R = I_o/I_r$ between the object and reference intensities is equal to 1.

The transmittance versus exposure $(t–E)$ curve for a holographic film typically looks like that sketched in Figure P6.1 with a linear portion between the exposures E_1 and E_2. To get a linear response, it is therefore advantageous to have the exposure lying between these values. Calculate R in terms of E_1 and E_2 when the whole linear portion of the $t–E$ curve is utilized.

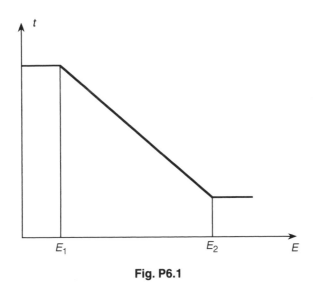

Fig. P6.1

6.4 A thin hologram is recorded with the object- and reference waves being unit amplitude, plane waves with angles of incidence θ_o and θ_r respectively, and with wavelength λ_1.
(a) Calculate the intensity distribution in the hologram plane.

The hologram is reconstructed with a plane wave with angle of incidence θ_i and wavelength λ_2.

(b) Use the grating equation (Equation (4.19)) to find an expression for the angle θ_s of the reconstructed wave and the ray angle magnification $M_\alpha \simeq \sin\theta_s/\sin\theta_o$.

6.5 As mentioned in Section 5.4.2, a thermoplastic film has a bandlimited spatial frequency response centred at about 1500 lines/mm. What should preferably be the angle between the object- and reference waves when using this film?

6.6 The transmittance of a phase hologram is given by Equation (6.10). Do you get a single reconstructed wave? What happens to the zero-order wave as M approaches 2.4?

Π6.7 (a) Consider the interferogram in Figure 6.8(a). If we have $\theta_1 = 10°, \theta_2 = 20°$ and an He–Ne laser is used, what is the maximum deflection of the plate?
(b) Assuming $g = 2$ and $\lambda = 632.8$ nm, what is the maximum vibration amplitude on the left and right side of the turbine blade in B of Figure 6.8(d) (Consult a table of Bessel functions)

6.8 Consider the vibrating bar in Figure 6.10a) and let the angular vibrating amplitude be α_m. Assume that we record a time-average hologram with a modulated reference wave with a vibrating amplitude of the reference mirror equal to R and the phase $\psi_r = \pi$. Sketch (qualitatively) the intensity distribution of the resulting interferogram.

7 Moiré technique

7.1 INTRODUCTION

Figure 3.2 is an illustration of two interfering plane waves. Let us look at the figure for what it really is, namely two gratings that lie in contact with a small angle between the grating lines. As a result, we see a fringe pattern of much lower frequency than the individual gratings. This is an example of the moiré effect and the resulting fringes are called moiré fringes or a moiré pattern. Figures 3.4, 3.7 and 3.8 are examples of the same effect. The mathematical description of moiré patterns resulting from superposition of sinusoidal gratings is the same as for interference patterns formed by electromagnetic waves. The moiré effect is therefore often termed mechanical interference. The main difference lies in the difference is wavelength which constitutes a factor of about 10^2 and greater.

The moiré effect can be observed in our everyday surroundings. Examples are: folded fine-meshed curtains (moiré means watered silk), rails on each side of a bridge or a staircase, nettings, etc.

On the TV screen it is often possible to observe moiré effects. For instance, when actors in striped suits are standing at a certain distance from the camera, moiré fringes are formed between the stripes and the lines of the TV image (see Section 5.7.3).

Moiré patterns can be formed by many types of gratings. Examples are: circular gratings (see Figure 3.7), radial gratings, circular zone plates, spirals, etc. In this way, artistic patterns can be created: they have become an art in themselves.

Often, gratings applied in moirè methods are transparencies with transmittances given by a square-wave function. Instead of square-wave functions, we describe linear gratings by sinusoidal transmittances (reflectances) bearing in mind that all types of periodic gratings can be described as a sum of sinusoidal gratings. A sinusoidal grating of constant frequency is given by

$$t_1(x, y) = a + a \cos\left(\frac{2\pi}{p} x\right) \qquad (7.1)$$

where p is the grating period and where $0 < a < \frac{1}{2}$. The principle behind measurement applications of gratings is that they in some way become phase modulated (see Section 4.7). This means that the grating given by Equation (7.1) can be expressed as

$$t_2(x, y) = a + a \cos 2\pi\left(\frac{x}{p} + \psi(x)\right) \qquad (7.2)$$

$\psi(x)$ is the modulation function and is equal to the displacement of the grating lines from its original position divided by the grating period

$$\psi(x) = \frac{u(x)}{p} \tag{7.3}$$

where $u(x)$ is the displacement.

When the two gratings given by Equations (7.1) and (7.2) are laid in contact, the resulting transmittance t becomes the product of the individual transmittances, viz.

$$t(x, y) = t_1 t_2$$

$$= a^2 \left\{ 1 + \cos\frac{2\pi}{p}x + \cos 2\pi\left[\frac{x}{p} + \psi(x)\right] + \tfrac{1}{2}\cos 2\pi\left[\frac{2x}{p} + \psi(x)\right] + \tfrac{1}{2}\cos 2\pi\psi(x) \right\} \tag{7.4}$$

The first three terms represent the original gratings, the fourth term the second grating with doubled frequency, while the fifth term depends on the modulation function only. It is this term which describes the moiré pattern.

Another way of combining gratings is by addition (or subtraction). This is achieved by e.g. imaging the two gratings by double exposure onto the same negative. By addition we get

$$t(x, y) = t_1 + t_2 = 2a\left\{ 1 + \cos \pi\psi(x)\cos 2\pi\left[\frac{x}{p} + \frac{1}{2}\psi(x)\right] \right\} \tag{7.5}$$

Here we see that the term $\cos \pi\psi(x)$ describing the moiré fringes are amplitude modulating the original grating.

Both Equations (7.4) and (7.5) have a maximum resulting in a bright fringe whenever

$$\psi(x) = n, \quad \text{for} \quad n = 0, \pm 1, \pm 2, \pm 3, \ldots \tag{7.6}$$

and minima (dark fringes) whenever

$$\psi(x) = n + \tfrac{1}{2}, \quad \text{for} \quad n = 0, \pm 1, \pm 2, \pm 3, \ldots \tag{7.7}$$

Both grating t_1 and t_2 could be phase-modulated by modulation functions ψ_1 and ψ_2 respectively. Then $\psi(x)$ in Equations (7.6) and (7.7) has to be replced by

$$\psi(x) = \psi_1(x) - \psi_2(x) \tag{7.8}$$

In both multiplication and addition (subtraction), the grating becomes demodulated (see Section 3.5.4) thereby getting a term depending solely on $\psi(x)$, describing the moiré fringes. By using square wave (or other types) of gratings, the result will be completely analogous.

Below we shall find the relations between $\psi(x)$ (and $u(x)$) and the measuring parameters for the different applications.

7.2   MOIRÉ BETWEEN TWO ANGULARLY DISPLACED GRATINGS

The mathematical description of this case is the same as for two plane waves interfering under an angle α (see Section 3.3). When two gratings of transmittances t_1 and t_2 are laid in contact, the resulting transmittance is not equal to the sum $t_1 + t_2$ as in Section 3.3, but the product $t_1 \cdot t_2$. The result is, however, essentially the same, i.e. the gratings form a moiré pattern with interfringe distance (cf. Equation (3.21))

$$d = \frac{p}{2\sin\dfrac{\alpha}{2}} \tag{7.9}$$

This can be applied for measuring α by measurement of d.

7.3 MEASUREMENT OF IN-PLANE DEFORMATION AND STRAINS

When measuring in-plane deformations a grating is attached to the test surface. When the surface is deformed, the grating will follow the deformation and will therefore be given by Equation (7.2). The deformation $u(x)$ will be given directly from Equation (7.3):

$$u(x) = p\psi(x) \tag{7.10}$$

To obtain the moiré pattern, one may apply one of several methods (Post 1982; Sciammarell 1972, 1982):

(1) Place the reference grating with transmittance t_1 in contact with the model grating with transmittance t_2. The resulting intensity distribution then becomes proportional to the product $t_1 \cdot t_2$.

(2) Image the reference grating t_1 onto the model grating t_2. The resulting intensity then becomes proportional to the sum $t_1 + t_2$. This can also be done by forming the reference grating by means of interference between two plane coherent waves.

(3) Image the model grating t_2, and place the reference grating t_1 in the image plane. t_1 then of course has to be scaled according to the image magnification. The resulting intensity becomes proportional to $t_1 \cdot t_2$.

(4) Image the reference grating given by t_1 onto a photographic film and thereafter image the model grating given by t_2 after deformation onto another film. Then the two films are laid in contact. The result is $t_1 \cdot t_2$.

(5) Do the same as under (4) except that t_1 and t_2 are imaged onto the same negative by double exposure. The result is $t_1 + t_2$.

Other arrangements might also be possible. In applying methods (1), (3) and (4), the resulting intensity distribution is proportional to $t_1 \cdot t_2$ and therefore given by Equation (7.4) which can be written

$$I(x) = I_0 + I_1 \cos 2\pi\psi(x) + \text{terms of higher frequencies} \tag{7.11}$$

By using methods (2) and (5), the intensity distribution becomes equal to $t_1 + t_2$ and therefore given by Equation (7.5), which can be written

$$I(x) = I_0 + I_1 \cos \pi\psi(x) \cos \frac{2\pi x}{p} + \text{other terms} \tag{7.12}$$

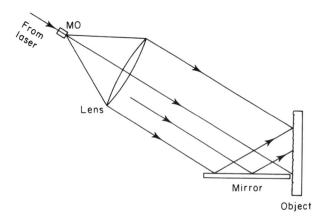

Fig. 7.1

We see that by using methods (1), (3) and (4) we essentially get a DC-term I_0, plus a term containing the modulation function. In methods (2) and (5) this last term amplitude modulates the original reference grating. When applying low-frequency gratings, all these methods may be sufficient for direct observation of the modulation function, i.e. the moiré fringes. When using high-frequency gratings, however, direct observation might be impossible due to the low contrast of the moiré fringes. This essentially means that the ratio I_1/I_0 in Equations (7.11) and (7.12) is very small. We then have the possibility of applying optical filtering (see Section 4.5). For methods (4) and (5), this can be accomplished by placing the negative into a standard optical filtering set-up. Optical filtering techniques can be incorporated directly into the set-up of methods (1) and (2) by using coherent light illumination and observing the moiré patterns in the first diffracted side orders. A particularly interesting method (belonging to method (2)) devised by Post (1971) is shown in Figure 7.1. Here the reference grating is formed by interference between a plane wave and its mirror image. The angle of incidence and grating period are adjusted so that the direction of the first diffracted side order coincides with the object surface normal. Experiments using model gratings of frequencies as high as 600 lines/mm have been reported by application of this method. To get sufficient amount of light into the first diffraction order one has to use phase-type gratings as the model grating. For the description of how to replicate fine diffraction gratings onto the object surface the reader is referred to the work of Post.

By using methods (3), (4) and (5) the grating frequency (i.e. the measuring sensitivity) is limited by the resolving power of the imaging lens. For curved surfaces, the model grating will be modulated due to the curvature, which can lead to false information about the deformation when using methods (1), (2) and (3). This is not the case for methods (4) and (5) because this modulation is incorporated in the reference grating (the first exposure). Surface curvature might also be a problem when using methods (3), (4) and (5) because of the limited depth of focus of the imaging lens. If we neglect the above-mentioned drawbacks, methods (1), (2) and (3) have the advantage of measuring the deformation in real time.

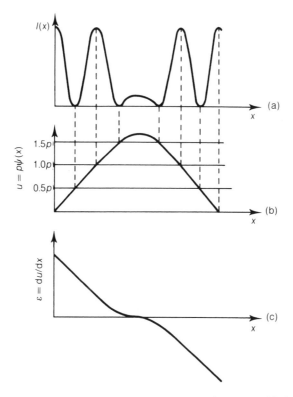

Fig. 7.2. *(a) Example of the intensity distribution of a moiré pattern with the corresponding (b) displacement and (c) strain*

By using one of these methods, we will, either directly or by means of optical filtering, obtain an intensity distribution of the same form as given in the two first terms in Equation (7.11) or (7.12). This distribution has a

maximum whenever $\psi(x) = n$, for $n = 0, 1, 2, \ldots$
minimum whenever $\psi(x) = n + \frac{1}{2}$, for $n = 0, 1, 2, \ldots$

According to Equation (7.10) this corresponds to a displacement equal to

$$u(x) = np \qquad \text{for} \quad \text{maxima} \qquad (7.13a)$$

$$u(x) = (n + \tfrac{1}{2})p \qquad \text{for} \quad \text{minima} \qquad (7.13b)$$

Figure 7.2(a) shows an example of such an intensity distribution with the corresponding displacement and strain in Figures 7.2(b) and (c).

By orienting the model grating and the reference grating along the y-axis, we can in the same manner find the modulation function $\psi_y(y)$ and the displacement $v(y)$ in the y-direction. $\psi_x(x)$ and $\psi_y(y)$ can be detected simultaneously by applying crossed gratings, i.e. gratings of orthogonal lines in the x- and y-directions. Thus we also are able to calculate the strains

$$\varepsilon_x = p \frac{\partial \psi_x}{\partial x} \tag{7.14a}$$

$$\varepsilon_y = p \frac{\partial \psi_y}{\partial y} \tag{7.14b}$$

$$\gamma_{xy} = p \left[\frac{\partial \psi_x}{\partial y} + \frac{\partial \psi_y}{\partial x} \right] \tag{7.14c}$$

7.3.1 Methods for increasing the sensitivity

In many cases the sensitivity, i.e. the displacement per moiré fringe, may be too small. A lot of effort has therefore been put into increasing the sensitivity of the different moiré techniques (Gåsvik and Fourney 1986). The various amendments made to the solution of this problem can be grouped into three methods: fringe multiplication, fringe interpolation and mismatch.

Fringe multiplication

In moiré methods one usually employs square-wave or phase gratings as model gratings. An analysis of such gratings would have resulted in expressions for the intensity distribution equivalent to Equations (7.11) and (7.12), but with an infinite number of terms containing frequencies which are integral multiples of the basic frequency. When using such gratings it is therefore possible to filter out one of the higher-order terms by means of optical filtering. By filtering out the Nth order, one obtains N times as many fringes and therefore an N-fold increase of the sensitivity compared to the standard technique. This is the concept of fringe multiplication. However, the intensity distribution of the harmonic terms generally decreases with increasing orders which therefore sets an upper bound to the multiplication process. Although in some special cases multiplications up to 30 have been reported, practical multiplications can rarely exceed 10.

Fringe interpolation

This method consists of determining fractional fringe orders. It can be done by scanning the fringe pattern with a slit detector or taking microdensitometer readings from a photograph of the fringes. It can also be done by digitizing the video signal from a TV picture. These methods are limited by the unavoidable noise in the moiré patterns. When forming the reference grating by interference between two plane waves, interpolation can be achieved by moving the phase of one of the plane waves. This is easily obtained by means of f.ex. a quarterwave plate and a rotatable polarizer in the beam of the plane wave.

For more details of such methods, see Chapter 11.

Mismatch

This is a term concerning many techniques. It consists of forming an initial moiré pattern between the model and reference grating before deformation. Instead of counting fringe orders due to the deformation, one measures the deviation or curvature of the initial pattern. The initial pattern can be produced by gratings having different frequencies, by a small rotation between the model and reference grating or by a small gap between them. In this way one can increase the sensitivity by at least a factor of ten.

This is equivalent to the spatial carrier method described in Section 11.4.3.

7.4 MEASUREMENT OF OUT-OF-PLANE DEFORMATIONS. CONTOURING

7.4.1 Shadow moiré

We shall now describe an effect where moiré fringes are formed between a grating and its own shadow: so-called shadow moiré. The principle of the method is shown in Figure 7.3.

The grating lying over the curved surface is illuminated under the angle of incidence θ_1 (measured from the grating normal) and viewed under an angle θ_2. From the figure we see that a point P_0 on the grating is projected to a point P_1 on the surface which by viewing is projected to the point P_2 on the grating. This is equivalent to a displcement of the grating relative to its shadow equal to

$$u = u_1 + u_2 = h(x, y)(\tan \theta_1 + \tan \theta_2) \tag{7.15}$$

where $h(x, y)$ is the height difference between the grating and the point P_1 on the surface. In accordance with Equation (7.3), this corresponds to a modulation function equal to

$$\psi(x) = \frac{u}{p} = \frac{h(x, y)}{p}(\tan \theta_1 + \tan \theta_2) \tag{7.16}$$

A bright fringe is obtained whenever $\psi(x) = n$, for $n = 0, 1, 2, \ldots$, which gives

$$h(x, y) = \frac{np}{\tan \theta_1 + \tan \theta_2} \tag{7.17a}$$

and

$$h(x, y) = \frac{(n + \frac{1}{2})p}{\tan \theta_1 + \tan \theta_2} \tag{7.17b}$$

for dark fringes. In this way, a topographic map is formed over the surface.

In the case of plane wave illumination and observation from infinity, θ_1 and θ_2 will remain constant across the surface and Equation (7.17) describes a contour map with

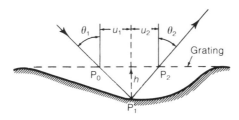

Fig. 7.3. *Shadow moiré*

a constant, fixed contour interval. With the point source and the viewing point at finite distances, θ_1 and θ_2 will vary across the surface resulting in a contour interval which is dependent on the surface coordinates. This is of course an unsatisfactory condition. However, if the point source and the viewing point are placed at equal heights z_p above the surface and if the surface height variations are negligible compared to z_p, then $\tan \theta_1 + \tan \theta_2$ will be constant across the surface resulting in a constant contour interval. This is a good solution, especially for large surface areas which are impossible to cover with a plane wave because of the limited aperture of the collimating lens.

If the surface height variations are large compared to the grating period, diffraction effects will occur, prohibiting a mere shadow of the grating to be cast on the surface. Shadow moiré is therefore best suited for rather coarse measurements on large surfaces. It is relatively simple to apply and the necessary equipment is quite inexpensive. It is a valuable tool in experimental mechanics and for measuring and controlling shapes.

Perhaps the most successful application of the shadow moiré method is in the area of medicine, such as the detection of scoliosis, a spinal disease which can be diagnozed by means of the asymmetry of the moiré fringes on the back of the body. Takasaki (1973, 1982) has worked extensively with shadow moiré for the measurement of the human body. He devised a grating made by stretching acrylic monofilament fibre on a frame using screws or pins as the pitch guide. According to him, the grating period should be 1.5–2.0 mm, and the diameter should be half the grating period. The grating should be sprayed black with high-quality dead back paint. Figure 7.4 shows an example of contouring of a mannequin of real size using shadow moiré.

7.4.2 Projected fringes

We now describe a method where fringes are projected onto the test surface (Idesawa *et al.* 1977; Takeda 1982; Doty 1983). Figure 7.5 shows fringes with an inter-fringe distance d projected onto the xy-plane under an angle θ_1 to the z-axis. The fringe period along the x-axis then becomes

$$d_x = \frac{d}{\cos \theta_1} \tag{7.18}$$

Also in the figure is drawn a curve S representing a surface to be contoured. From the figure we see that a fringe originally positioned at P_1 will be displaced to P_2. This displacement is given by

$$u = z(\tan \theta_1 + \tan \theta_2) \tag{7.19}$$

where z is the height of P_2 above the xy-plane and θ_2 is the viewing angle. From Equation (7.3) this gives a modulation function equal to

$$\psi(x) = \frac{u}{d_x} = \frac{z(\tan \theta_1 + \tan \theta_2)}{d/\cos \theta_1} = \frac{z}{d}(\sin \theta_1 + \cos \theta_1 \tan \theta_2)$$

$$= \frac{z}{d}\frac{\sin(\theta_1 + \theta_2)}{\cos \theta_2} = \frac{z}{d}G \tag{7.20}$$

Fig. 7.4. *Shadow moiré contouring. (Reproduced from Takasaki 1973 by permission of Optical Society of America.)*

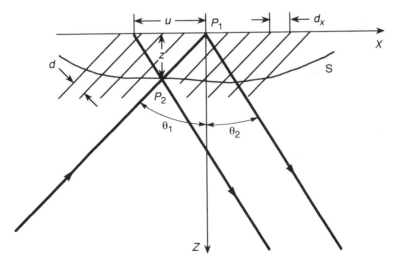

Fig. 7.5. *Fringe projection geometry.* θ_1 = *projection angle.* θ_2 = *viewing angle*

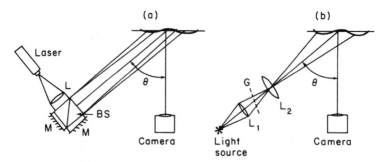

Fig. 7.6. *Fringe projection by means of (a) interference, (b) grating imaging. L = lenses, M = mirrors, BS = beamsplitter, G = grating*

where we have introduced the geometry factor

$$G = G(\theta_1, \theta_2) = \sin\theta_1 + \cos\theta_1 \tan\theta_2 = \frac{\sin(\theta_1 + \theta_2)}{\cos\theta_2} \qquad (7.21)$$

One method of projecting fringes on a surface is by means of interference between two plane waves inclined a small angle α to each other (Brooks and Heflinger 1969). This can be achieved by means of a Twyman–Green interferometer with a small tilt of one of the mirrors (Figure 7.6(a)). The distance between the interference fringes will then be equal to

$$d = \frac{\lambda}{2\sin(\alpha/2)} \qquad (7.22)$$

where λ is the wavelength and α is the angle between the two plane waves.

From Equation (7.2) and Equations (7.18–7.21), the intensity distribution across the surface then can be written as

$$I = 2\left\{1 + \cos 2\pi\left[\frac{x}{d_x} + \psi(x)\right]\right\} = 2\left\{1 + \cos\frac{2\pi}{d}[x\cos\theta_1 + zG]\right\} \qquad (7.23)$$

Figure 7.6(b) shows another method for projecting a fringe pattern onto the surface. Here, a grating is imaged onto the surface by means of lens L_2 (Pirodda 1982; Haliona *et al.* 1993; Yatagai and Idesawa 1982; Reid *et al.* 1984; Cline *et al.* 1984; Varman 1984; Dessus and Leblanc 1973; Gåsvik 1983; Cardenas–Garcia *et al.* 1994). The spatial distance d between the fringes is now given as

$$d = m_p p \qquad (7.24)$$

where p is the grating period and m_p is the magnification of the projecting system (lens L_2).

By imaging two surfaces with modulation functions

$$\psi_1 = \frac{z_1}{d}G \qquad (7.25a)$$

and

$$\psi_2 = \frac{z_2}{d}G \qquad (7.25b)$$

we get, according to Equations (7.6) and (7.7) (either by addition or subtraction), bright

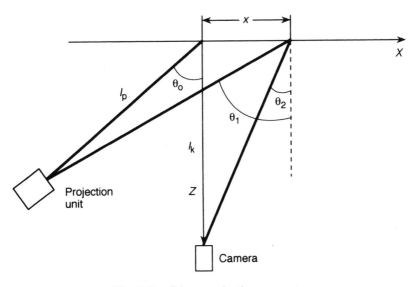

Fig. 7.7. *Fringe projection geometry*

fringes whenever

$$z_1 - z_2 = \frac{nd}{G} \quad \text{for} \quad n = 0, \pm 1, \pm 2, \ldots \quad (7.26a)$$

and dark fringes whenever

$$z_1 - z_2 = (n + \tfrac{1}{2})\frac{d}{G}, \quad \text{for} \quad n = 0, \pm 1, \pm 2, \ldots \quad (7.26b)$$

We will therefore observe an image of the object superposed by contour lines in the same way as the contour lines on a topographic map where the contour interval is given by Equation (7.26). In this way we can map the deformation of an object where the contour lines are referred to the undeformed surface, or we get a map of the topography of a surface referred to another surface which can be displayed in a 360° view (Asundi *et al.* 1994).

When the camera is pointing along the z-axis, we see from Figure 7.7 that

$$\tan \theta_1 = \frac{l_p \sin \theta_0 + x}{l_p \cos \theta_0} \quad (7.27a)$$

$$\tan \theta_2 = \frac{-x}{l_k} \quad (7.27b)$$

where θ_0 is the projection angle measured from the z-axis and l_p and l_k are the projection and camera distances respectively.

The dependence of u and ψ on the x-coordinate then becomes

$$u(x) = z(\tan \theta_1 + \tan \theta_2) = z\left[\frac{l_p \sin \theta_0 + x}{l_p \cos \theta_0} - \frac{x}{l_k} \right] = z\left[\tan \theta_0 + \frac{(l_k - l_p \cos \theta_0)x}{l_p l_k \cos \theta_0} \right] \quad (7.28a)$$

$$\psi(x) = \frac{z}{d}(\tan \theta_1 + \tan \theta_2)\cos \theta_1 = \frac{z}{d}\left[\sin \theta_0 + \frac{(l_k - l_p \cos \theta_0)x}{l_p l_k} \right]\frac{l_p}{\sqrt{l_p^2 + 2l_p x \sin \theta_0 + x^2}}$$

$$(7.28b)$$

From Equation (7.28a) we see that the displacement u becomes dependent on x only through z (i.e. the contour interval becomes independent on the position on the surface) if the projection lens and the camera lens are placed at equal heights above the xy-plane ($l_k - l_p \cos \theta_0 = 0$).

7.4.3 Vibration analysis

In the same manner as for holographic interferometry, moiré technique using projected fringes (or shadow moiré) can be applied for vibration analysis of surfaces (Hazell and Niven 1968; Vest and Sweeney 1972; Harding and Harris 1983). Let us analyse this method more closely.

Assume that a point on the surface in Figure 7.5 executes harmonic out-of-plane vibrations given by

$$z = z_0 + a \cos \omega t \tag{7.29}$$

where z_0 is the equilibrium position, a is the amplitude and ω the frequency.

The intensity distribution of the projected pattern (cf. Equation (7.23)) now becomes

$$I(x, t) = 2 \left[1 + \cos \frac{2\pi}{d} (x \cos \theta + (z_0 + a \cos \omega t) \sin \theta) \right] \tag{7.30}$$

where we for simplicity have assumed that the camera is imaging from infinity along the z-axis, i.e. $G = \sin \theta$ where θ is the projection angle.

The expression can be written as

$$I(x, t) = 2[1 + \cos(\phi_c + \phi_t)] = 2 + e^{i(\phi_c + \phi_t)} + e^{i(\phi_c + \phi_t)} \tag{7.31}$$

where

$$\phi_c = (2\pi/d)(x \cos \theta + z_0 \sin \theta) \tag{7.32a}$$

$$\phi_t = (2\pi/d) \sin \theta \, a \cos \omega t \tag{7.32b}$$

By photographing this pattern with an exposure time much longer than the vibration period T, the resulting transmittance t of the film becomes proportional to $I(x, t)$ averaged over the vibration period. This is analogous to time-average holography (see Section 6.7) and gives for the transmittance

$$t = \frac{1}{T} \int_0^T I(x, y) \, dt = 2 + \frac{e^{i\phi_c}}{T} \int_0^T e^{i\phi_t} \, dt + \frac{e^{-i\phi_c}}{T} \int_0^T e^{-i\phi_t} \, dt \tag{7.33}$$

Now we have

$$\frac{1}{T} \int_0^T e^{\pm i\phi_t} \, dt = \frac{1}{T} \int_0^T \exp[\pm i(2\pi/d) \sin \theta \, a \cos \omega t] \, dt$$

$$= J_0 \left(\frac{2\pi}{d} a \sin \theta \right) \tag{7.34}$$

which inserted into Equation (7.33) gives

$$t = 2 \left[1 + J_0 \left(\frac{2\pi}{d} a \sin \theta \right) \cos \phi_c \right] \tag{7.35}$$

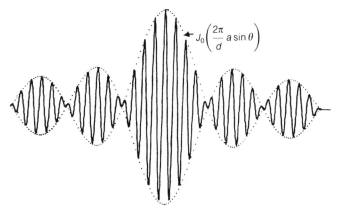

Fig. 7.8

where J_0 is the zeroth-order Bessel function. From this expression we see that the Bessel function amplitude modulates the fringe pattern on the static surface given by z_0. This is illustrated in Figure 7.8.

The negative with the transmittance given from Equation (7.35) can be placed in an optical filtering system in the same way as in the static case, resulting in an amplitude distribution in the image plane equal to

$$u = J_0\left(\frac{2\pi}{d} a \sin \theta\right) e^{i\phi_c} \tag{7.36}$$

and hence an intensity distribution given by

$$I = J_0^2\left(\frac{2\pi}{d} a \sin \theta\right) \tag{7.37}$$

From the values of the arguments of the Bessel function corresponding to its maximum and zeros given on page 145, we find that light fringes occur when

$$a = \frac{d}{2\pi \sin \theta} \times [0, 3.83, 7.02, 10.17, \ldots] \tag{7.38a}$$

and dark fringes occur when

$$a = \frac{d}{2\pi \sin \theta} \times [2.40, 5.52, 8.65, 11.79, \ldots] \tag{7.38b}$$

The first dark fringe of this pattern thus corresponds to an amplitude a_l equal to

$$a_l = 0.38 \frac{d}{\sin \theta} \tag{7.39}$$

which is a figure representing the sensitivity of the method.

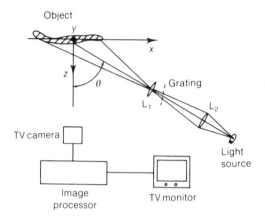

Fig. 7.9. *Projection moiré using TV-camera and digital image processor*

7.4.4 Moiré technique by means of digital image processing

A very convenient way of adding (subtracting) pictures is by means of digital image processing (Gåsvik 1983). A set-up for projection moiré is shown in Figure 7.9 where a grating is projected onto the object. The surface with projected fringes are imaged by a TV camera and the video signal is sent to an image processor, see Chapter 10. In this way it is possible to subtract a stored image from the image seen by the camera in real time. Figure 7.10 shows some examples of the results obtained with such a system. Figure 7.10(a) shows a cartridge casing with a dent. The reference image stored into the memory is taken from the undefective side of the casing. Figure 7.10(b) shows the result from two recordings of a 25-litre oil can before and after filling with water. In Figure 7.10(c) the system is applied to vibration analysis. It shows a circular plate centrally clamped to a shaft and excited by a shaker at a point in the lower right edge. The picture is a time-average recording resulting in a zeroth-order Bessel fringe function displaying the amplitude distribution of the plate as described in Section 7.4.3. For time-average recordings, the image processor is not necessary.

7.5 REFLECTION MOIRÉ

As we have already seen, moiré technique offers a wide variety of application methods. Most of them are, however, variations of the basic principles discussed in the preceding sections. Here we mention a method which to a certain extent differs from the previous techniques. It is called Lichtenberg's method (Lichtenberg 1955) and can be applied on shiny, mirror-like surfaces and phase-objects (Liasi and North 1994).

Figure 7.11 shows the principle of the method. The smoothness of the surface S makes it possible to image the mirror image of the grating G by means of the lens L. As in previous methods, a grating can be placed in the image plane of L or the mirror image of G can be photographed before and after the deformation of S. The result is a moiré pattern defining the derivative of the height profile, i.e. the slope of the deformation.

In an analysis of the resolution of the reflection moiré method it is found that the maximum resolution that can be obtained with a viewing camera is of the order 7×10^{-3} radians.

Fig. 7.10. *Examples of TV-moiré fringes: (a) cartridge casing with a dent. Contour interval 0.15 mm; (b) 25-litre oil can after filling with water; (c) time-average recording of a 400-mm diameter aluminium plate excited in the lower right edge. Frequency 250 Hz. Amplitude corresponding to the first dark fringe 0.16 mm*

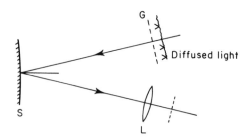

Fig. 7.11.   *Reflection moiré*

PROBLEMS

7.1 Two gratings with amplitude transmittance

$$t(x, y) = a\left[1 + \cos\left(\frac{2\pi}{p} x\right)\right]$$

are laid in contact with an angle α between the grating lines. Calculate $t_1 \cdot t_2$.

7.2 A circular zone plate with centre at (x_0, y_0) has an amplitude transmittance given by

$$t(x, y) = \tfrac{1}{2}\{1 + \cos \beta[(x - x_0)^2 + (y - y_0)^2]\}$$

where β is a constant. Suppose that two zone plate transmittances are laid in contact with a displacement d between their centres.

(a) Show that the resulting moiré fringes describes a new zone-plate pattern.
(b) Find the centre and the frequency of this new zone plate pattern.

7.3 We attach a grating of period p to a bar, whereafter the bar undergoes a uniaxial tension resulting in a strain equal to ε_x.

(a) Write down the transmittances t_1 and t_2 of the grating before and after the load.
(b) Find an expression for $t_1 + t_2$ describing the moiré pattern.
(c) If $\varepsilon_x = 10^{-3}$, the grating frequency is 20 lines/mm and the length of the bar is 10 cm, how many moiré fringes are observed?

7.4 In shadow moiré, the reference grating t_1 is multiplied by the object (shadow) grating t_2. In what sense can we say that this is a sort of synchronous demodulation of t_2? What is the low-pass filter in this case?

7.5 Consider the method of projected fringes using interference between two plane waves (Figure 7.5(a)). Assume that the camera is looking normal to the surface from infinity (i.e. $\theta_2 = 0$).

(a) calculate the out-of-plane displacement Δz per fringe in this case.
(b) Based on pure geometric considerations, there is a limit to the sensitivity (displacement per fringe) in this set-up. Find this limit and the corresponding values of θ_1 and α (the angle between the two plane waves).
(c) Using $\lambda = 600$ nm, calculate the mean spatial frequency in lines/mm on the surface in this case. Why is this measuring sensitivity unattainable?

8 Speckle methods

8.1 THE SPECKLE EFFECT

In Figure 8.1, light is incident on, and scattered from, a rough surface of height variations greater than the wavelength λ of the light. As is shown in the figure, light is scattered in all directions. These scattered waves interfere and form an interference pattern consisting of dark and bright spots or speckles which are randomly distributed in space. In white light illumination, this effect is scarcely observable owing to lack of spatial and temporal coherence (see Section 3.2). Applying laser light, however, gives the scattered light a characteristic granular appearance as shown in the image of a speckle pattern in Figure 8.2.

It is easily realized that the light field at a specific point in a speckle pattern must be the sum of a large number N of components representing the light from all points on the scattering surface. The complex amplitude at point in a speckle pattern can therefore be written as

$$u = \frac{1}{\sqrt{N}} \sum_{k=1}^{N} u_k = \frac{1}{\sqrt{N}} \sum_{k=1}^{N} U_k e^{i\phi_k} \qquad (8.1)$$

By assuming that (1) the amplitude and phase of each component are statistically independent and also independent of the amplitudes and phases of all other components, and (2) the phases ϕ_k are uniformly distributed over all values between $-\pi$ and $+\pi$, Goodman (1975) has shown that the complex amplitude u will obey Gaussian statistics. Further he has shown that the probability density function P_1 for the intensity in a speckle

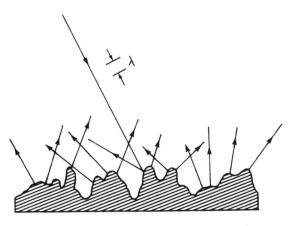

Fig. 8.1. *Light scattering from a rough surface*

Fig. 8.2.   *Photograph of a speckle pattern*

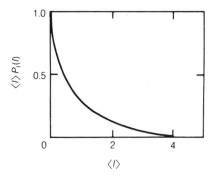

Fig. 8.3

pattern is given as

$$P_1(I) = \frac{1}{\langle I \rangle} \exp\left(-\frac{1}{\langle I \rangle} \right) \tag{8.2}$$

where $\langle I \rangle$ is the mean intensity. The intensity of a speckle pattern thus obeys negative exponential statistics. Figure 8.3 shows a plot of $P_1(I)$. From this plot we see that the most probable intensity value is zero, that is, black.

A measure of the contrast in a speckle pattern is the ratio $C = \sigma_1/\langle I \rangle$, where σ_1 is the standard deviation of the intensity given by

$$\sigma_1^2 = \langle \Delta I^2 \rangle = \langle (I - \langle I \rangle)^2 \rangle = \langle I^2 - 2\langle I \rangle I + \langle I \rangle^2 \rangle = \langle I^2 \rangle - \langle I \rangle^2 \tag{8.3}$$

where the brackets denote mean values. By using

$$\langle I^2 \rangle = \int_0^\infty P_1(I) I^2 \, dI = 2\langle I \rangle^2 \tag{8.4}$$

we find the contrast C in a speckle pattern to be unity.

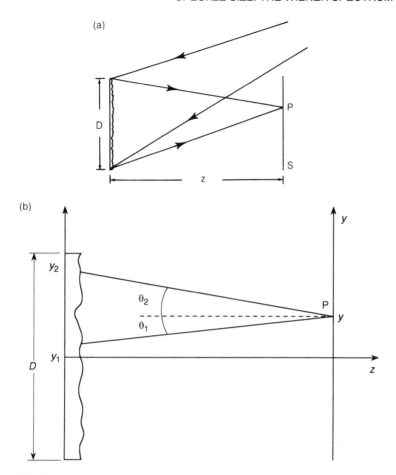

Fig. 8.4. *(a) Objective speckle formation; (b) Speckle formation at the point P(y) from two points of coordinates y_1 and y_2*

8.2 SPECKLE SIZE. THE WIENER SPECTRUM

From Figure 8.2 we see that the size of the bright and dark spots varies. To find a representative value of the speckle size, consider Figure 8.4(a), where a rough surface is illuminated by laser light over an area of cross-section D. The resulting so-called objective speckle pattern is observed on a screen S at a distance z from the scattering surface. For simplicity, we consider only the y-dependence of the intensity. An arbitrary point P on the screen will receive light contributions from all points on the scattering surface. Let us assume that the speckle pattern at P is a superposition of the fringe patterns formed by light scattered from all point pairs on the surface. Any two points separated by a distance l will give rise to fringes of frequency $f = l/(\lambda z)$ (see Section 3.5.1, Equation (3.32)). The fringes of highest spatial frequency f_{max} will be formed by the two edge points, for which

$$f_{max} = \frac{D}{\lambda z} \tag{8.5}$$

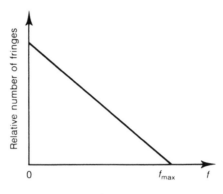

Fig. 8.5

The period of this pattern is a measure of the smallest objective speckle size σ_o which therefore is

$$\sigma_o = \frac{\lambda z}{D} \qquad (8.6)$$

For smaller separations l, there will be a large number of point pairs giving rise to fringes of the corresponding frequency. The number of point pairs separated by l is proportional to $(D-l)$. Since the various fringe patterns have random individual phases they will add incoherently. The contribution of each frequency to the total intensity will therefore be proportional to the corresponding number of pairs of scattering points. Since this number is proportional to $(D-l)$, which in turn is proportional to $(f_{max} - f)$, the relative number of fringes versus frequency, i.e. the spatial frequency spectrum will be linear, as shown in Figure 8.5.

A more formal (one-dimensional) treatment of this phenomenon goes as follows: In Figure 8.4(b) we consider the contribution to the field at P from two points on the surface with coordinates y_1 and y_2. The rays from y_1 and y_2 to P are making the incidence angles θ_1 and θ_2 respectively. From Equation (1.9b), the fields from y_1 and y_2 at P therefore can be written as ($z = 0$):

$$u_1 = A(y_1)e^{ik_1y} = U(y_1)e^{i\phi_1}e^{ik_1y}$$
$$u_2 = A(y_2)e^{ik_2y} = U(y_2)e^{i\phi_2}e^{ik_2y} \qquad (8.7a)$$

where

$$A(y_1) = U(y_1)e^{i\phi_1}$$
$$A(y_2) = U(y_2)e^{i\phi_2} \qquad (8.7b)$$

and

$$k_{1,2} = k \sin\theta_{1,2} = \frac{2\pi}{\lambda}\left(\frac{y_{1,2} - y}{z}\right) \qquad (8.7c)$$

where we by the last equality have introduced the paraxial approximation, and where ϕ_1 and ϕ_2 are random spatial phases due to the random height variations of the surface. The intensity thus becomes

$$I(y) = U^2(y_1) + U^2(y_2) + A(y_1)A^*(y_2)e^{i(k_1-k_2)y} + A^*(y_1)A(y_2)e^{-i(k_1-k_2)y}$$

$$= I(y_1) + I(y_2) + 2U(y_1)U(y_2)\cos\left[\frac{2\pi}{\lambda z}(y_2 - y_1)y + \phi_{12}\right] \qquad (8.8)$$

where $\phi_{12} = \phi_2 - \phi_1$. We therefore see that the light scattered from the points y_1 and y_2 forms a sinusoidal fringe pattern at P of spatial frequency

$$f_{12} = \frac{y_2 - y_1}{\lambda z} = \frac{s_{12}}{\lambda z} \qquad (8.9)$$

with random phase ϕ_{12}. Equation (8.8) can therefore be interpreted as

$$I = \langle I \rangle + \Delta I \qquad (8.10)$$

where $\langle I \rangle = I_m = I(y_1) + I(y_2)$ is the mean intensity and ΔI is the fluctuating interference term.

There are many pairs of points on the object separated by the spacing s_{12}. All such pairs contribute a fringe component of the same frequency f_{12}. These many fringes at frequency f_{12} add with random spatial phase ϕ_{12}.

The Fourier spectrum of Equation (8.8) is given as

$$\mathscr{F}\{I(y)\} = [I(y_1) + I(y_2)]\delta(f_y) + A(y_1)A^*(y_2)\delta(f_y + f_{12}) + A^*(y_1)A(y_2)\delta(f_y - f_{12}) \qquad (8.11)$$

where

$$\delta(f_y \pm f_{12}) = \delta\left(f_y \pm \frac{y_2 - y_1}{\lambda z}\right) = \pm(\lambda z)\delta(y_2 - y_1 \pm \lambda z f_y) \qquad (8.12)$$

If we square the amplitude of each of these Fourier components, and average over an ensemble of microscopically different objects, the resulting number is the value of the power spectrum at frequency f_y. When squaring and averaging $I(y)$ of Equation (8.8), we must, however, evoke the fact that from Equation (8.3) we have

$$\langle I^2 \rangle = \langle I \rangle^2 + \langle \Delta I^2 \rangle \qquad (8.13)$$

The averaging we do by integrating y_1 and y_2 over the surface from $-D/2$ to $D/2$. Since $I(y_1), I(y_2)$ is zero outside the surface, we can change the integration limits to $-\infty, \infty$. Since y_1 and y_2 are interchangeable, we must divide the result by 2. The power spectrum $W(f_y)$, also termed the Wiener spectrum, is therefore given by

$$2W(f_y) = \delta(f_y)\left\{\int\int_{-\infty}^{\infty}[I(y_1) + I(y_2)]\,dy_1\,dy_2\right\}^2$$

$$+ (\lambda z)^2\int_{-\infty}^{\infty}I(y_1)\,dy_1\int_{-\infty}^{\infty}I(y_2)\delta(y_2 - y_1 - \lambda z f_y)\,dy_2$$

$$+ (\lambda z)^2\int_{-\infty}^{\infty}I(y_1)\,dy_1\int_{-\infty}^{\infty}I(y_2 - y_1 + \lambda z f_y)\,dy_2$$

$$= 2I_m^2\delta(f_y) + (\lambda z)^2\int_{-\infty}^{\infty}I(y_1)I(y_1 - \lambda z f_y)\,dy_1 + (\lambda z)^2\int_{-\infty}^{\infty}I(y_1)I(y_1 + \lambda z f_y)\,dy_1 \qquad (8.14)$$

By a simple change of variable in the last integral it becomes equal to the second last integral, and therefore

$$W(f_y) = I_m^2 \delta(f_y) + (\lambda z)^2 \int_{-\infty}^{\infty} I(\eta)I(\eta - \lambda z f_y)\, d\eta \tag{8.15}$$

where we have changed the integration variable. This integral is recognized as the autocorrelation of $I(x)$. An obvious generalization to two dimensions for an isotropic surface finally gives for the Wiener spectrum $W(f_x, f_y)$:

$$W(f_x, f_y) = I_m^2 \left\{ \delta(f_x, f_y) + \frac{\displaystyle\int_{-\infty}^{\infty} I(\xi, \eta)I(\xi - \lambda z f_x, \eta - \lambda z f_y)\, d\xi\, d\eta}{\left| \displaystyle\int_{-\infty}^{\infty} I(\xi, \eta)\, d\xi\, d\eta \right|^2} \right\} \tag{8.16}$$

As an example, consider the case of a uniformly bright square object of width D:

$$I(x, y) = I_m \operatorname{rect}\left(\frac{x}{D}\right) \operatorname{rect}\left(\frac{y}{D}\right) \tag{8.17}$$

we get for the Wiener spectrum

$$W(f_x, f_y) = I_m^2 \left\{ \delta(f_x, f_y) + \left(\frac{\lambda z}{D}\right)^2 \Lambda\left(\frac{\lambda z}{D} f_x\right) \Lambda\left(\frac{\lambda z}{D} f_y\right) \right\} \tag{8.18}$$

where Λ is the triangle function. This calculation is completely analogous to the calculation of the optical transfer function for a square aperture, see Section 4.6.2 and gives the same result as in Figure 8.5.

Figure 8.6 shows the same situation as in Figure 8.4 except that the scattering surface now is imaged on to a screen by means of a lens L. The calculation of the size of the resulting so-called subjective speckles is analogous to the calculation of the objective speckle size. Here the cross-section of the illuminated area has to be exchanged by the diameter of the imaging lens. The subjective speckle size σ_s therefore is given as

$$\sigma_s = \frac{\lambda b}{D} \tag{8.19}$$

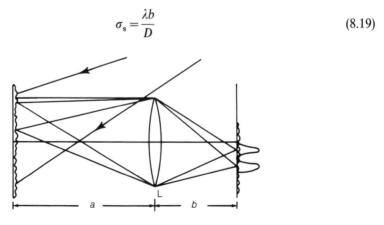

Fig. 8.6. *Subjective speckle formation*

where b is the image distance. By introducing the aperture number

$$F = \frac{f}{D} \qquad (8.20)$$

where f is the focal length, we get

$$\sigma_s = (1 + m)\lambda F \qquad (8.21)$$

where $m = (b - f)/f$ is the magnification of the imaging system. From this equation we see that the speckle size increases with decreasing aperture (increasing aperture number). This can be easily verified by stopping down the eye aperture when looking at a speckle pattern.

Speckle formation in imaging cannot be explained by means of geometrical optics which predicts that a point in the object is imaged to a point in the image. The field at a point in the image plane therefore should receive contributions only from the conjugate object point, thus preventing the interference with light from other points on the object surface. However, even an ideal lens will not image a point into a point but merely form a intensity distribution (the Airy disc, see Section 4.6) around the geometrical image point due to diffraction of the lens aperture. This is indicated in Figure 8.6. It is therefore possible for contributions from various points on the object to interfere so as to form a speckle pattern in the image plane.

8.3 SPECKLE PHOTOGRAPHY

Discussions on the subject of speckle photography can be found in Burch and Tokarski (1968), Archbold *et al.* (1970), Archbold and Ennos (1972), Duffy (1972, 1974), Cloud (1975), Dainty (1975), Erf (1978), Fourney (1978), Hung (1978), Tiziani (1972), Jones and Wykes (1979) and Halliwell and Pickering (1993).

8.3.1 The Fourier fringe method

Assume that we image a speckle pattern onto a photographic film. After development this negative is placed in the object plane of a set-up for optical filtering like that in Figure 4.12. Figure 8.7 shows (Fourney 1978)

(1) a speckle pattern on the negative;
(2) the resulting diffraction pattern (the spatial frequency spectrum) in the x_f, y_f-plane; and
(3) typical form of the smoothed intensity distribution along the x_f-axis.

The dark spot in the middle of Figure 8.7(b) is due to the blocking of the strong zeroth-order component. Figure 8.7(c) displays the essential feature discussed in Section 8.2, namely that the imaged speckle pattern contains a continuum of spatial frequencies ranging from zero to $f_{max} = \pm 1/\sigma_s$, where σ_s is the smallest speckle size given from Equation (8.19). For a circular aperture the frequency distribution will not be exactly linear as in Figure 8.5, but merely as indicated in Figure 8.7(c).

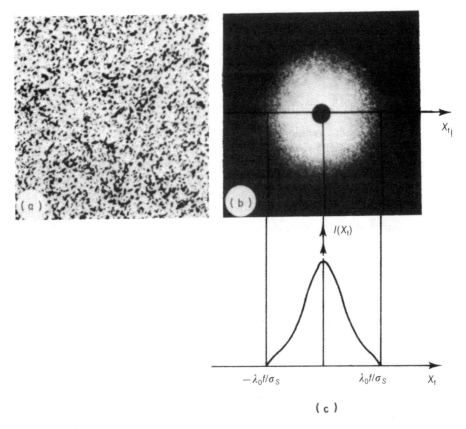

Fig. 8.7. *(a) Speckle pattern and its corresponding (b) diffraction pattern and (c) intensity distribution along the x_1-axis. ((a) and (b) reproduced from Fourney (1978) by permission of Academic Press, New York.)*

Now assume that we in the diffraction plane (the filter plane) place a screen with a hole a distance x_f from the optical axis. This situation is illustrated in Figure 8.8, where Figure 8.8(a, b) shows the spectra before and after the filtering process has taken place. The same spectrum (as in Figure 8.8(b)) would have resulted by filtering out the first side order from a sinusoidal grating of frequency

$$f_x = \frac{x_f}{\lambda_0 f} \tag{8.22}$$

where f is the focal length of the transforming lens and λ_0 is the wavelength of the light source applied in the filtering process. What we have actually done therefore is to select one of the numerous gratings with a continuum of directions and frequencies that constitute the laser-illuminated object.

As a consequence, we can imagine that a grating given by

$$t_1 = a(1 + \cos 2\pi f_x x) \tag{8.23}$$

is attached to the object surface. When the object undergoes an in-plane deformation, the

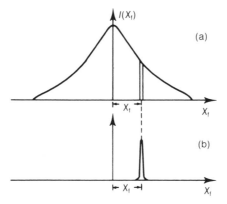

Fig. 8.8. *Spectrum of a speckle pattern (a) before and (b) after filtering*

speckle pattern will follow the displacements of the points on the object surface. Consequently, the grating will be phase-modulated and thus can be written as

$$t_2 = a[1 + \cos 2\pi(f_x x + \psi_x(x))] \tag{8.24}$$

This is closely analogous to the situation described in Section 7.3 where a model grating was attached to the object for measurement of an in-plane deformation by means of moiré technique. In the same way as described there, we can image the speckle pattern before and after the deformation onto the same film negative by double exposure. The resulting transmittance then becomes equal to $t_1 + t_2$, the sum of Equations (8.23) and (8.24). The two speckle patterns could possibly also be imaged onto two separate negatives and subsequently superposed, giving a resultant transmittance equal to $t_1 \cdot t_2$, but this is more difficult to achieve. By means of optical filtering of the double-exposed negative we get an intensity distribution dependent on the modulation function only (see Section 4.7, Equation (4.64)). This distribution will be maximum for $\psi_x(x) = n$ and minimum for $\psi_x(x) = n + (1/2)$ and correspond to displacements (cf. Equations (7.10) and (7.13)) equal to

$$u(x) = n \frac{1}{f_x} = n \frac{\lambda_0 f}{x_f} \qquad \text{for maxima} \tag{8.25a}$$

$$u(x) = (n + \tfrac{1}{2}) \frac{\lambda_0 f}{x_f} \qquad \text{for minima} \tag{8.25b}$$

in direct analogy with the moiré method. In deriving Equation (8.25) we have assumed unit magnification. These values therefore must be divided by the magnification m applied when recording the speckle patterns.

The speckle pattern represents gratings of all orientations in the plane of the object (cf. the spectrum in Figure 8.7(b)). Therefore, by placing the hole in the screen in the filter plane a distance y_f along the y_f-axis we obtain the modulation function $\psi_y(y)$ and the corresponding displacement $v_y(y)$ along the y-axis. Generally, by placing the hole in the filter plane at a point of coordinates x_f, y_f, the resulting intensity distribution will give

a displacement equal to

$$s(x, y) = \sqrt{u^2(x) + v^2(y)} = n\lambda_0 f \sqrt{(1/x_f)^2 + (1/y_f)^2} \qquad \text{for maxima} \qquad (8.26a)$$

$$s(x, y) = (n + \tfrac{1}{2})\lambda_0 f \sqrt{(1/x_f)^2 + (1/y_f)^2} \qquad \text{for minima} \qquad (8.26b)$$

In this case we thus measure the displacement along a direction inclined an angle β to the x-axis given by

$$\tan \beta = y_f/x_f \qquad (8.27)$$

In this method we can therefore vary the measuring sensitivity by varying x_f and y_f, and a remarkable property of this technique is that the sensitivity can be varied subsequent to the actual measurement, that is, after the speckle patterns have been recorded. Highest sensitivity is obtained by placing the filtering hole at the edge of the spectrum. The first-order dark fringe then corresponds to a displacement equal to

$$s_{\min} = \frac{1}{2}\sigma_s = \frac{1}{2}\frac{\lambda b}{D} = 0.5(1 + m)\lambda F \qquad (8.28)$$

With a magnification $m = 1$, this gives a sensitivity equal to the laser wavelength multiplied by the aperture number of the imaging lens. The sensitivity limit by using laser speckle photography can therefore be down to about 1 μm.

Figure 8.9 shows an example of such Fourier fringes obtained by this method (Hung 1978). Note that as the filtering hole is moved away from the optical axis, the number of fringes increases, i.e. the sensitivity is increased.

Fig. 8.9. *Fringe patterns depicting the horizontal and vertical displacements of a cantilever beam obtained from the various filtering positions in the Fourier filtering plane. (Reproduced from Hung 1978 by permission of Academic Press, New York.)*

8.3.2 The Young fringe method

On photographing a speckle pattern, each speckle will form a pointlike blackening on the film. When the object undergoes an in-plane deformation, the speckle pattern will follow this deformation. A double-exposed negative of two speckle patterns resulting from a deformation therefore will consist of identical point pairs separated by a distance equal to the deformation times the magnification of the imaging system.

Assume that we illuminate this double-exposed negative with an unexpanded laser beam. When the beam covers one pair of identical points, they will act in the same way as the two holes P_1 and P_2 in the screen of the Young's interferometer (see Figure 3.12, Section 3.5.1). On a screen at a distance z from the negative we therefore will observe interference fringes which are parallel and equidistant and with a direction perpendicular to the line joining P_1 and P_2, i.e. perpendicular to the displacement.

The situation is sketched in Figure 8.10. In Section 3.5.1 we found that the distance between two adjacent fringes in this pattern is equal to

$$d = \frac{\lambda z}{D} \tag{8.29}$$

where D is the distance between P_1 and P_2. If the displacement on the object is equal to s the separation of the corresponding speckle points on the negative is equal to $m \cdot s$ where m is the magnification of the camera. Put into Equation (8.29) this gives, for the object deformation,

$$s = \frac{\lambda z}{md} \tag{8.30}$$

In deriving the equations for the Young fringe pattern in Section 3.5.1 we used the approximation $\sin \theta = \tan \theta$. Without this approximation Equation (8.30) becomes

$$s = \frac{\lambda}{m \sin \theta} \tag{8.31}$$

By measuring the fringe separation d we can therefore find the object displacement at the point of the laser beam incidence using Equation (8.30). Better accuracy is obtained by measuring the distance d_n covered by n fringes on the screen. We then have $d = d_n/n$,

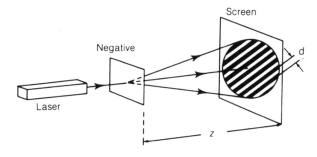

Fig. 8.10. *Young fringe formation*

which gives

$$s = \frac{n\lambda z}{md_n}$$

(8.32)

To obtain such a Young fringe pattern, the identical pairs of speckles must be separated by a distance which is at least equal to one half of the speckle size, that is $1/2\sigma_s$. This is the same sensitivity limit as found in Equation (8.28).

Fig. 8.11. *Young fringes at different points in a plate under tension in a miniature rig*

By placing the film negative into a slide, movable in both the horizontal and vertical directions, measurement on different points on the film, i.e. the object surface, is performed quite easily and quickly. This is, however, in principle a pointwise measurement in contrast to the Fourier fringe method, which is a full field measurement. On the other hand, the Young fringe method gives better accuracy and the fringes are more easily obtained. If the Young's fringes do not appear from a double exposed specklegram it is, under normal circumstances, also impossible to obtain the Fourier fringes.

Figure 8.11 shows an example of the results obtained by this method. The object is a metal plate under tension in a miniature rig.

8.4 SPECKLE INTERFEROMETRY

Reports from early work on speckle interferometry can be found in Leendertz (1979), Groh (1968), Hung and Hovanesian (1972), and Butters and Leendertz (1971).

Figure 8.12 shows two plane waves u_1 and u_2 symmetrically incident on an object surface at the angles $\pm \theta$ to the z-axis. The object is imaged by a camera directed along the z-axis. u_1 and u_2 will be scattered at the object surface and form two separate speckle patterns that interfere in the film plane of the camera. Consider an arbitrary point (x, y) on the object surface. This point is imaged onto the film plane where it receives contributions due to u_1 and u_2 with a certain mutual phase difference. When the object is deformed, this phase difference will change. For an out-of-plane displacement (along the z-axis) or a displacement along the y-axis, this phase change will be zero. For a displacement d along the x-axis, we see from the figure that the resulting phase change becomes equal to

$$\Delta\phi = k2d \sin \theta \qquad (8.33)$$

By double exposure of the object before and after deformation followed by optical filtering

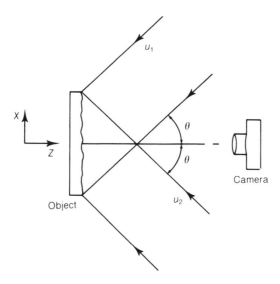

Fig. 8.12. *Speckle interferometry*

of the negative one obtains a fringe pattern that gives the in-plane deformation equal to $\lambda/(2 \sin \theta)$ per fringe.

A mathematical treatment of this process can be given as follows.

Assume that the field amplitude distribution over the object surface due to a normally incident plane wave of unit amplitude is given by $u_0(x, y)$. The two obliquely incident plane waves will be given by $u_1 = e^{ikx \sin \theta}$ and $u_2 = e^{-ikx \sin \theta}$ which give, for the field amplitude distribution over the object surface in the first recording,

$$u_{01} = u_0(x, y)(e^{ikx \sin \theta} + e^{-ikx \sin \theta}) \tag{8.34}$$

In Section 4.6 we found that the relation between the image (u_i) and object (u) amplitude distribution is given by a convolution integral

$$u_i(x_i, y_i) = \int_{-\infty}^{\infty} \int hu(x, y) \, dx \, dy \tag{8.35}$$

where $h = h(x_i - mx, y_i - my)$ is the impulse response of the imaging lens, m is the magnification and x_i, y_i are the spatial coordinates in the image plane. In the first exposure, the amplitude distribution in the image plane therefore is given by

$$u_{i1} = u_{1+} + u_{1-} = U_{1+} e^{i\phi_{1+}} + U_{1-} e^{i\phi_{1-}} \tag{8.36}$$

where

$$u_{1\pm} = \int_{-\infty}^{\infty} \int hu_0(x, y) e^{\pm ikx \sin \theta} \, dx \, dy \tag{8.37}$$

After a displacement d in the x-direction the amplitude distribution over the object surface becomes

$$u_{02} = u_0(x + d, y)(e^{ik(x+d) \sin \theta} + e^{-ik(x+d) \sin \theta}) \tag{8.38}$$

This gives for the field amplitude in the image plane in the second exposure

$$u_{i2} = u_{2+} + u_{2-} \tag{8.39}$$

where

$$u_{2\pm} = e^{\pm ikd \sin \theta} \int_{-\infty}^{\infty} \int hu_0(x + d, y) e^{\pm ikx \sin \theta} \, dx \, dy \tag{8.40}$$

For a displacement less than the speckle size we can put to a good approximation $u_0(x + d, y) = u_0(x, y)$ which inserted into Equation (8.40) gives

$$u_{2\pm} = e^{\pm ikd \sin \theta} u_{1\pm} \tag{8.41}$$

The intensities I_1 and I_2 recorded in the first and second exposures therefore can be written as

$$I_1 = |u_{1+} + u_{1-}|^2 = U_{1+}^2 + U_{1-}^2 + 2U_{1+} U_{1-} \cos \delta \tag{8.42a}$$

$$I_2 = |u_{2+} + u_{2-}|^2 = |u_{1+} e^{ikd \sin \theta} + u_{1-} e^{-ikd \sin \theta}|^2$$
$$= U_{1+}^2 + U_{1-}^2 + 2U_{1+} U_{1-} \cos(\delta + 2kd \sin \theta) \tag{8.42b}$$

where $\delta = \phi_{1+} - \phi_{1-}$. We see that the combined speckle pattern becomes phase

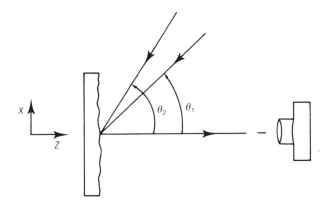

Fig. 8.13. *Non-symmetrical incidence*

modulated in the second exposure. The amplitude transmittance of this double-exposed negative becomes proportional to $I_1 + I_2$. On placing this negative into an optical filtering set-up and blocking out the zero-order and all negative (or positive) frequencies, the complex amplitude distribution in the image plane becomes

$$u_f = U_{1+} U_{1-} [e^{i\delta}(1 + e^{i2kd\sin\theta})] \tag{8.43}$$

and the intensity

$$I_f = 2(U_{1+} U_{1-})^2 [1 + \cos(2kd\sin\theta)] \tag{8.44}$$

This distribution has a maximum whenever

$$2kd\sin\theta = n2\pi \tag{8.45}$$

i.e. when $d = n\,(\lambda/2\sin\theta)$ as given previously.

A high displacement sensitivity can therefore be obtained by this method. This sensitivity is reduced by decreasing the angle θ, but the displacement cannot be greater than the speckle size which therefore sets a lower limit on the sensitivity. The reason for this is seen from our mathematical analysis. In this case we are not allowed to put $u_0(x + d, y) = u_0(x, y)$ and Equation (8.41) is no longer valid. The random phases of the speckle pattern will then also modulate the displacement fringes after filtering, resulting in a fringe pattern completely buried in speckle noise.

By using non-symmetrical incidence of the two plane waves as shown in Figure 8.13, the method also becomes sensitive to out-of-plane displacements (Hung 1978). In this case the relative phase change between the two speckle wavefronts due to a deformation is given by

$$\Delta\phi = k[(\cos\theta_1 - \cos\theta_2)w + (\sin\theta_1 - \sin\theta_2)u] \tag{8.46}$$

where w is the out-of-plane displacement.

Speckle interferometry demands the same stability requirements as in holographic interferometry. An advantage is that in-plane deformations can be independently measured and further that the fringes always localize on the object surface. High-resolution film is not required since it is only necessary to resolve the speckles, the size of which is proportional to the aperture number of the imaging lens. A major drawback of speckle interferometry is the poor quality of the displacement fringes and that the object area to be measured is limited by the size of the collimating lenses forming the incident plane waves.

8.5   SPECKLE-SHEARING INTERFEROMETRY

Laser speckle methods can be utilized in many different ways. One method, the speckle-shearing interferometer, is particularly interesting because it enables direct measurements of displacement derivatives which are related to the strains (Hung and Taylor 1973; Leendertz and Butters 1973; Hung (1974, 1982), Hung *et al.* 1975).

The principle of speckle-shearing interferometry, also termed shearography, is to bring the rays scattered from one point of the object into interference with those from a neighbouring point. This can be obtained in a speckle-shearing interferometric camera as depicted in Figure 8.14. The set-up is the same as that used in ordinary speckle photography, except that one half of the camera lens is covered by a thin glass wedge. In that way, the two images focused by each half of the lens are laterally sheared with respect to each other. If the wedge is so oriented that the shearing is in the x-direction, the rays from a point $P(x, y)$ on the object will interfere in the image plane with those from a neighbouring point $P(x + \delta x, y)$. The shearing δx is proportional to the wedge angle.

When the object is deformed there is a relative displacement between the two points that produces a relative optical phase change $\Delta \phi$ given by

$$\Delta \phi = k\{(1 + \cos \theta)[w(x + \delta x, y) - w(x, y)] \\ + \sin \theta[u(x + \delta x, y) - u(x, y)]\} \qquad (8.47)$$

where θ is the angle of incidence and u and w are the displacement components in the x- and z-directions, respectively. If the shear δx is small, the relative displacements may be approximated by the displacement derivatives and thus Equation (8.47) becomes

$$\Delta \phi = k\left[(1 + \cos \theta)\frac{\partial w}{\partial x} + \sin \theta \frac{\partial u}{\partial x}\right]\delta x \qquad (8.48)$$

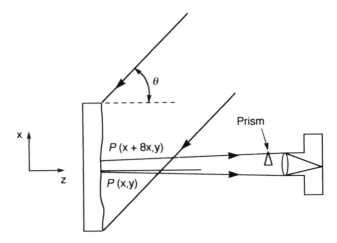

Fig. 8.14.   *Speckle-shearing interferometric camera*

By rotating the camera $90°$ about the z-axis, u in Equation (8.48) is replaced by v, the displacement component in the y-direction.

By double exposure of the object before and after deformation, a speckle fringe pattern depicting $\Delta\phi$ of Equation (8.48) will be generated. This pattern will be clearly visible by placing the negative in an optical filtering set-up and blocking out the zero order component. Dark fringes occur when

$$\Delta\phi = (2n + 1)\pi, \quad \text{for } n = 0, 1, 2 \ldots \tag{8.49}$$

Equation (8.48) shows that to measure $\partial w/\partial x$ only, one may employ normal illumination ($\theta = 0°$), which gives

$$\Delta\phi = \frac{4\pi}{\lambda}\frac{\partial w}{\partial x}\delta x \tag{8.50}$$

but it is not possible to isolate $\partial u/\partial x$. This can be obtained by recording two fringe patterns using two different illumination angles. Then $\partial u/\partial x$ can be pointwise separated from $\partial w/\partial x$.

Compared to speckle interferometry, shearography has a lot of advantages. Good quality fringes are easily obtained and the vibration isolation and the coherence length requirements are greatly reduced. Like holography, shearography has been adapted to electronic cameras combined with digital image processing. A convenient means for obtaining the shear is to place a Michelson interferometer set-up with a tilt of one of the mirrors in front of the image sensor. By an ingenious set-up constructed by the group at Luleå University of Technology simultaneous observation of the out-of-plane displacement and slope in real time is made possible. The set up is shown in Figure 8.15 where an electronic holography configuration is combined with a shearography con-

Fig. 8.15. *Schematic for simultaneous measurement of out-of-plane displacement and slope. (M, mirror; BS, beam splitter). (From Saldner, H.O. (1994) Electronic holography and Shearography in experimental mechanics. Licentiate thesis, Lulea University of Technology.)*

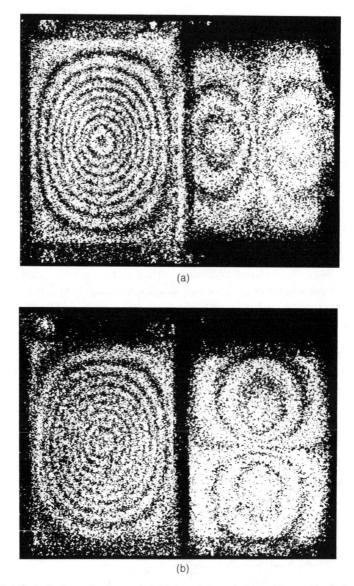

(a)

(b)

Fig. 8.16. *(a) Out-of-plane displacement fringes (w) and slope fringes (δw/δx) as seen from the monitor for a rectangular aluminium plate clamped along its edges and loaded at the centre. The shear Δx is 6 mm, and the displacement w = 2.5 μm. (b) Out-of-plane displacement fringe pattern (w) and slope pattern (δw/δy) for the same object as in (a). The shear Δy is 7 mm. (From Mohan, N.K., Saldner, H. and Molin, N.E. (1993) Electronic speckle pattern interferometry for simultaneous measurements of out-of-plane displacement and slope, Opt letters, **18**, 1861–3. Reproduced by courtesy of The Optical Society of America and of H. Saldner and N.E. Molin*

figuration so that the displacement and slope interferograms are displayed in each half of the monitor. The result of an experiment (Mohan *et al.* 1993) conducted on a rectangular aluminium plate of dimensions 65 mm × 100 mm × 0.6 mm is shown in Figure 8.16.

(a)

(b)

Fig. 8.17. *Electronic shearography (ES) used for non-destructive testing of a ceramic material. (a) A vertical crack is clearly visualized by ES as a fringe in the centre of the sample. (b) The crack is not detected using TV holography. ((From Saldner, H.O. (1994) Electronic holography and Shearography in experimental mechanics. Licentiate thesis, Lulea University of Technology.)*

Figure 8.17 shows another advantage of shearography. Figure 8.17(a) shows the phase map of a ceramic object using shearography, while Figure 8.17(b) is a phase map of the same object using holography (Saldner 1994). A crack in the centre of the object becomes visible in the upper image due to the fact that shearography is highly sensitive to local variations in the deformation field. In the holographic phase map of Figure 8.17(b), however, it is difficult or impossible to see the same crack.

Finally we mention that speckle-shearing interferometry can be obtained by intentional misfocusing of the object. Then it is possible for rays from neighbouring points on the object to interfere in the image plane.

8.6   WHITE-LIGHT SPECKLE PHOTOGRAPHY

The term 'white-light speckle photography' has been introduced because this technique utilizes the same principles as laser speckle photography. But here a white light source is used instead of a laser. The 'speckles' in this method are some type of structure attached to the object surface. A common technique is to cover the object surface with retroreflective paint. This paint consists of small glass spheres embedded in the emulsion and produces reflected points of light which are very bright when photographed from a position close to the illumination direction. These spheres represent a random distribution of points containing all spatial frequencies up to the reciprocal of the smallest sphere separation. This highest spatial frequency is therefore limited by the size of the glass spheres. However, if the spheres are small enough, or photographed from a sufficient distance, the highest spatial frequency will be limited by the impulse response of the imaging lens, i.e. the diameter of the Airy disc (see Section 4.6) in the same way as in laser speckle photography. There may be several ways of creating a random pattern on the object surface. One method is to spray the object with black paint on a white background or vice versa.

The working principle of white light speckle photography is the same as for laser speckle photography. Both the Fourier fringe and the Young fringe methods can be applied using the same equations in evaluating the patterns. The main difference is that the detailed structure of the white light speckles is fixed on the surface, not in space as with laser speckles. Consequently the requirement for high-quality recording optics is more severe. The random pattern on the surface must be resolved by the camera which requires a large aperture lens with minimum aberration. On the other hand, if the surface is curved, a good depth of focus is required to resolve the pattern over the whole object area. The depth of focus, however, varies inversely as the square of the lens aperture (see Section 4.6). These conflicting requirements are partly overcome by an ingenious modification of the camera made by Burch and Forno (Burch and Forno 1975; Forno 1975). They have mounted a mask with symmetrically disposed parallel slots inside the camera lens. In that way they tune the camera response to a particular spatial frequency determined by the slot separation. Thereby both the resolution and the depth of focus is increased for this tuned frequency.

White light speckle photography is easily applied in industrial environments. Figure 8.18 shows a white light speckle camera assembly. It consists of an ordinary photographic camera with a standard 50 mm lens mounted on a plate together with an electronic flash and a beamsplitter. This is so arranged to make the retroreflected light from the painted object surface fall inside the lens aperture. Figure 8.19 shows an example of the application of this camera. It is a 3 m high concrete tube of 0.75 m diameter and 50 mm tube wall thickness. Between the two exposures the tube has been horizontally loaded by 5000 kg with a jack at the upper right end. The picture shows Young fringe readings at some selected points on the negative. The displacement per fringe is 0.22 mm and results have been obtained over an object depth of about 30 cm.

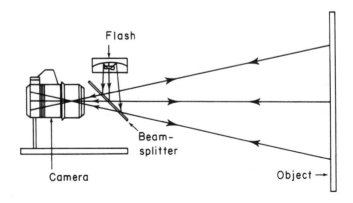

Fig. 8.18. *White light speckle camera*

Fig. 8.19. *Young fringes obtained by white light speckle photography on a 3-metre high concrete tube loaded at the upper right end*

(a)

Fig. 8.20. *(a) Notched Plexiglas® plate under tension in miniature rig. The resulting Fourier fringe patterns displaying the resulting (b) vertical and (c) horizontal displacements*

(b)

(c)

Fig. 8.20. *(b) and (c)*

In the same way as for laser speckles, we can talk of white-light subjective and objective speckles. To obtain the latter, we have to place the film in contact with the object surface. Figure 8.20 shows the results of such an experiment. Figure 8.20(a) shows the object, a notched plate made of Plexiglas loaded in tension in a miniature rig. A film plate was clamped to one side of the Plexiglas® plate which was rubbed to get speckles. The plate was illuminated from the other side with an ordinary lamp and loaded between the two exposures. Figure 8.20(b, c) shows the Fourier fringe pattern displaying the vertical and horizontal displacements obtained from this experiment.

PROBLEMS

8.1 In speckle photography we image a surface onto a film using a camera with aperture number F and a magnification of m. Assume the recorded intensity distribution (using a He–Ne laser) in the first recording is given by $I(x, y)$. Before the second recording the surface is displaced in the x-direction a distance s. After development the film is placed in the optical filtering set-up (using an He–Ne laser) shown in Figure 4.14 a distance z from the filter plane.
(a) Find (to an arbitrary multiplicative constant) the intensity distribution in the filter plane.
(b) The fringe period in the filter plane is d. Find s.
(c) For $d = 2\,\text{mm}$, $m = 0.1$ and $z = 100\,\text{mm}$, calculate s.
(d) What is the minimum measurable displacement s when $F = 4$, and when $F = 22$.

8.2 Suppose we are applying the white-light speckle method on an object with a transversal magnification m of our imaging optics. The sensitivity limit for the in-plane deformation is given by the Rayleigh criterion

$$s_{\min} = 1.22\lambda(1 + m)F$$

where F is the aperture number. The depth of focus for the lens is given by Equation (4.58), and the longitudinal magnification $m_{\text{L}} = m^2$. We assume that we obtain the sensitivity limit $s_{\min}$ also for object points which lies within the depth of focus given by Equation (4.58). (a) What is the sensitivity limit when we wish to measure on an object with depth variations up to Δa?

8.3 When imaging a speckle pattern a mask consisting of a double slit as sketched in Figure P4.4 is placed in the plane of the exit pupil.
Calculate the resulting Wiener spectrum $W(f_x, 0)$.

9 Photoelasticity and polarized light

9.1 POLARIZED LIGHT

Up to now we have treated the light field amplitude as a scalar quantity. The electromagnetic field, however, is a vector quantity which is perpendicular to the direction of propagation and with a defined orientation in space. This property is known as the polarizaton of light and is utilized in the field of photoelasticity. Before going into the details of this technique, we therefore take a brief survey of the fundamentals of polarized light.

Consider a plane wave propagating in the z-directon. The field amplitude is a vector denoted by

$$\mathbf{u} = \mathbf{U} e^{ikz} \tag{9.1}$$

of length U at an arbitrary angle ϕ to the x-axis, see Figure 9.1. The components along the x- and y-axis therefore are

$$U_x = U \cos \phi \tag{9.2a}$$

$$U_y = U \sin \phi \tag{9.2b}$$

We then have

$$\mathbf{u} = |U_x e^{i\delta_x} \mathbf{e}_x + U_y e^{i\delta_y} \mathbf{e}_y] e^{ikz} \tag{9.3}$$

where $\mathbf{e}_x$ and $\mathbf{e}_y$ are the unit vectors along the x- and y-axis respectively. By including the phase constants δ_x and δ_y we have taken care of the fact that the two components may not have the same phase. The factor e^{ikz} we omit, since it merely gives the orientation of the z-axis. Equation (9.3) then can be written as

$$\mathbf{u} = e^{i\beta/2}[U_x e^{-i\delta/2} \mathbf{e}_x + U_y e^{i\delta/2} \mathbf{e}_y]$$
$$= e^{i\beta/2}[u_x \mathbf{e}_x + u_y \mathbf{e}_y] \tag{9.4}$$

where

$$\beta = \delta_x + \delta_y \tag{9.5a}$$

$$\delta = \delta_y - \delta_x \tag{9.5b}$$

$e^{i\beta/2}$ is a common phase factor which can be omitted since it does not affect the orientation of $\mathbf{u}$. This orientation, termed the direction of polarization or the state of polarization, is therefore completely fixed by two independent quantities, i.e.

$$\tan \phi = U_y/U_x$$
$$\text{and} \quad \delta \tag{9.6}$$

Fig. 9.1

The intensity of this wave is given as

$$I = |u|^2 = U_x^2 + U_y^2 \tag{9.7}$$

It is evident and easy to prove that the intensity always becomes equal to the sum of the squares of the field components in a Cartesian coordinate system independent of the orientation of the coordinate axes.

When such a polarized wave passes a fictional plane perpendicular to the z-axis, the tip of the U-vector will in the general case describe an ellipse in that plane. A general state of polarization is therefore termed 'elliptically polarized light'. When $\delta = \pm \pi/2$ and $U_x = U_y$, the ellipse degenerates to a circle and when $\delta = 0$ or π to a straight line. These two special cases therefore represent so-called circular and plane (or linear) polarized light respectively (see Figure 9.2). More specifically, $\delta = +\pi/2$ represents right circularly polarized light, whilst $\delta = -\pi/2$ represents left circularly polarized light. This has to do with the direction of rotation of the tip of the field vector.

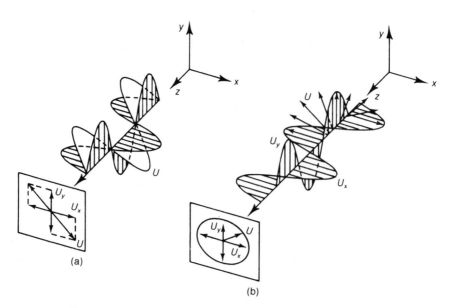

(a)

(b)

Fig. 9.2. *(a) Linear polarized light. (b) Circularly polarized light*

9.2 POLARIZING FILTERS

To alter and to analyse the state of polarization of the light, one places in the beam different types of polarization filters. Below, we consider the most important types.

9.2.1 The linear polarizer

A linear or plane polarizer (often called simply a polarizer) has the property of transmitting light which field vector is parallel to the transmission direction of the polarizer only. The field u_t transmitted through the polarizer therefore becomes equal to the component of the incident field onto this direction. If the incident wave U is plane-polarized at an angle α to the transmission direction of the polarizer, we therefore have

$$u_t = U \cos \alpha \tag{9.8}$$

and the intensity

$$I_t = |u_t|^2 = U^2 \cos^2 \alpha \tag{9.9}$$

Equation (9.9) is called Malus' law.

This is the case of an ideal polarizer with amplitude transmittances $t_1 = 1$ and $t_2 = 0$ parallel to an normal to the transmission direction respectively. In a real polarizer, $t_1 \leqslant 1$ and $t_2 \geqslant 0$, but $t_1 \gg t_2$. Consider two such identical linear polarizers placed in tandem in a light beam as shown in Figure 9.3. Here polarizer 1 has its transmission direction along the x-axis while the transmission axis of polarizer 2 is inclined an angle α to the x-axis. Assume that the wave field u_0 incident on polarizer 1 is linearly polarized at an angle ϕ to the x-axis with amplitude equal to 1.
Then

$$\mathbf{u}_0 = \cos \phi \mathbf{e}_x + \sin \phi \mathbf{e}_y \tag{9.10}$$

After passing polarizer 1, the field amplitude becomes

$$\mathbf{u}_1 = t_1 \cos \phi \mathbf{e}_x + t_2 \sin \phi \mathbf{e}_y \tag{9.11}$$

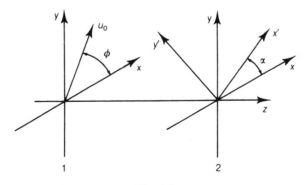

Fig. 9.3

To calculate the effect of polarizer 2, we first decompose $\mathbf{u}_1$ into the coordinate system of unit vectors $\mathbf{e}_{x'}$, $\mathbf{e}_{y'}$, along and normal to the transmission axis of polarizer 2 respectively:

$$\mathbf{u}_1 = (t_1 \cos \phi \cos \alpha + t_2 \sin \phi \sin \alpha)\mathbf{e}_{x'} + (-t_1 \cos \phi \sin \alpha + t_2 \sin \phi \cos \alpha)\mathbf{e}_{y'} \quad (9.12)$$

After passing polarizer 2, the field amplitude becomes

$$\mathbf{u}_2 = t_1(t_1 \cos \phi \cos \alpha + t_2 \sin \phi \sin \alpha)\mathbf{e}_{x'} + t_2(-t_1 \cos \phi \sin \alpha + t_2 \sin \phi \cos \alpha)\mathbf{e}_{y'} \quad (9.13)$$

and the intensity

$$I = |\mathbf{u}_2|^2 = (t_1^4 \cos^2 \phi + t_2^4 \sin^2 \phi) \cos^2 \phi$$
$$+ 2t_1 t_2 (t_1^2 - t_2^2) \sin \phi \cos \phi \sin \alpha \cos \alpha + t_1^2 t_2^2 \sin^2 \alpha \quad (9.14)$$

If the incident light is unpolarized (see section 9.3), all polarization angles ϕ from 0 to π will be equally represented. To find the transmitted intensity I_{up} in this case, we therefore have to average Equation (9.14) over this range of ϕ, i.e.

$$I_{up} = \frac{1}{\pi} \int_0^\pi I \, d\phi = (1/2)(T_1^2 + T_2^2)\cos^2 \alpha + T_1 T_2 \sin^2 \alpha \quad (9.15)$$

where we have substituted the intensity transmittances $T_1 = t_1^2$ and $T_2 = t_2^2$ for the corresponding amplitude transmittances of the polarizer. Equation (9.15) can be regarded as the generalized Malus law and can be used to determine T_1 and T_2 from measuring the intensity at, for example $\alpha = 0°$ and $90°$. Normally, T_1 and T_2 are wavelength dependent and the ratio T_2/T_1 is a measure of the quality of a linear polarizer. Values of T_2/T_1 as low as $10^{-5} - 10^{-6}$ can be reached with high-quality crystal polarizers.

A linear polarizer can be pictured as a grid of finely spaced parallel conducting metal threads. When the field of the incident wave oscillates parallel to the threads, it will induce currents in the threads. The energy of the field is therefore converted into electric current, which is converted into heat and the incident wave is absorbed. Because of the non-conducting spacings of the grid, currents cannot flow perpendicularly to the threads. Fields oscillating in the latter direction will therefore not produce any current and the light is transmitted. Although easy to understand, this type of polarizer is difficult to fabricate.

The most widely used linear polarizer is the polaroid sheet-type polarizer invented by E. H. Land. It can be regarded as the chemical version of the metal thread grating. Instead of long, thin threads it consists of long, thin molecules, i.e. long chains of polymeric molecules that contain many iodine atoms. These long molecules are oriented almost completely parallel to each other and because of the conductivity of the iodine atoms, the electric field oscillating parallel to the molecules will be strongly absorbed.

Another type of linear polarizer is made from double-refracting crystals. Double-refracton is a phenomenon where the incident light is split into two orthogonally linearly polarized components. By proper cutting and cementing of such crystals, usually calcite, one of the components is isolated and the other is transmitted, thereby giving a linear polarizer. Other phenomena utilized for the construction of linear polarizers are reflection and scattering.

9.2.2 Retarders

A retarder (or phase plate) is a polarization element with two orthogonal, characteristic directions termed the principal axes. When light passes a retarder, the field components parallel to the principal axes will acquire different phase contributions resulting in a relative phase difference δ. This is due to the phenomenon of double-refraction or birefringence which means that light fields oscillating parallel to the two principal axes 'experience' different indexes of refraction n_1 and n_2. A retarder plate of thickness t therefore produces a phase difference (retardance) equal to

$$\delta = \delta_2 - \delta_1 = k(n_2 - n_1)t \tag{9.16}$$

Figure 9.4 sketches a retarder with the principal axes h_1 and h_2 parallel to the x- and y-axis respectively. Consider a wave field U linearly polarized at an angle α to the x-axis normally incident on the retarder. The transmitted wave u_t then becomes

$$\mathbf{u}_t = U(\cos \alpha e^{i\delta_1}\mathbf{e}_x + \sin \alpha e^{i\delta_2}\mathbf{e}_y)$$
$$= Ue^{i(\delta_1 + \delta_2)/2}(\cos \alpha e^{-i\delta/2}\mathbf{e}_x + \sin \alpha e^{i\delta/2}\mathbf{e}_y) \tag{9.17}$$

Note that the intensity $|u_t|^2 = U^2$ is unchanged by passing the retarder. In the case of

$$\delta = \pm \pi/2 \tag{9.18}$$

and $\cos \alpha = \sin \alpha$; i.e. $\alpha = 45°$, u_t becomes circularly polarized.

A retarder with $\delta = \pm \pi/2$ is called a quarterwave plate because it corresponds to an optical path length difference equal to $\lambda/4$ (cf. Equation (9.16)). The combination of a linear polarizer followed by a quarterwave plate with their axes inclined at $45°$ therefore is termed a circular polarizer.

By putting $\delta = \pi$, Equation (9.17) becomes

$$\mathbf{u}_t = -iUe^{i(\delta_1 + \delta_2)/2}(\cos \alpha \mathbf{e}_x - \sin \alpha \mathbf{e}_y) \tag{9.19}$$

We see that the outcoming wave is linearly polarized and that the field vector is the mirror image of the incoming field vector about the x-axis, i.e. the polarizaion angle has changed by 2α. A retarder with $\delta = \pm \pi$ is called a halfwave plate because it corresponds to an optical path-length difference equal to $\lambda/2$. A halfwave plate therefore offers a convenient means for rotating the polarization angle of a linearly polarized light wave by turning the axes of the halfwave plate by the desired amount.

A retarder is usually made from double-refracting crystals such as quartz or mica. In contrast to the construction of linear polarizers, where one of the doubly refracting components is isolated, both components are transmitted collinearly by proper cutting

Fig. 9.4

Fig. 9.5. *Babinet–Soleil compensator*

and orientation of the crystal. A retarder can also be made from stretched sheets of polyvinyl alcohol (PVA) in the same way as polaroids. In fact, cellophane sheets can be used as a retarder.

In the retarders mentioned so far, the retardance δ is fixed by the plate thickness. It would be highly desirable, however, to have a retarder in which the retardance could be continuously varied. This can be done by a device called a Babinet–Soleil compensator shown in Figure 9.5. Here two retarders of the same crystal of thicknesses t_1 and t_2 with their axes inclined at 90° are mounted together. The total retardance of the unit therefore becomes proportional to the thickness difference $t_1 - t_2$. The upper plate consists of two wedges, one of which can be moved relative to the other, thereby varying the effective thickness t_1. This movement is controlled by a micrometer screw. The result is a retarder with variable retardance which is uniform over the whole field of the compensator.

9.3 UNPOLARIZED LIGHT

Light from ordinary light sources is said to be unpolarized. This is something of a misnomer, but it means that the instantaneous direction of polarization will vary rapidly and randomly in time between 0 and 2π. A common way of illustrating unpolarized light propagating in the z-direction is as illustrated in Figure 9.6. This picture could be misleading; a better illustration would be a collection of ellipses of random orientations and eccentricities.

Mathematically, unpolarized light can be represented in terms of two arbitrary, orthogonal, linearly polarized waves of equal amplitudes for which the relative phase difference varies rapidly and randomly. This aspect is strongly related to the coherence properties of the light (see Section 3.2). By analogy with the degree of coherence, one also speaks about the degree of polarizaton.

We shall not here go into any further details of this phenomenon. It should, however, be easy to realize that unpolarized light will (1) be unaffected with regard to intensity when

Fig. 9.6. *Unpolarized light*

transmitted through a retarder, and (2) become linearly polarized by transmitting a linear polarizer, but with an intensity independent of the transmission direction of the polarizer.

9.4 REFLECTION AND REFRACTION AT AN INTERFACE

When light is incident at an interface between two media of different refractive indices, both the reflected and the transmitted light will in general change its state of polarization relative to the state of polarization of the incident light.

Consider Figure 9.7 where the light is incident from a medium of refractive index n_1, on a medium of refractive index n_2. The incident light field is decomposed into the components u_{ip} and u_{in} parallel and normal to the plane of incidence respectively. (The plane of incidence is defined as the plane spanned by the surface normal at the point of incidence and the incident light ray.) The corresponding quantities of the reflected light are denoted u_{rp} and u_{rn} and their amplitudes U are related by

$$U_{rp} = r_p U_{ip} \tag{9.20a}$$

$$U_{rn} = r_n U_{in} \tag{9.20b}$$

where r_p and r_n are the amplitude reflection coefficients. These so-called Fresnel reflection coefficients are given as

$$r_p = \frac{\tan(\theta_1 - \theta_2)}{\tan(\theta_1 + \theta_2)} \tag{9.21a}$$

$$r_n = \frac{-\sin(\theta - \theta_2)}{\sin(\theta_1 + \theta_2)} \tag{9.21b}$$

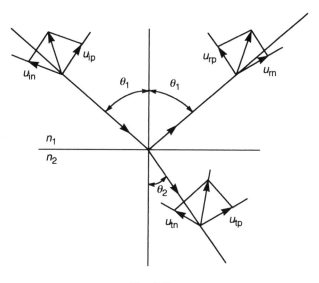

Fig. 9.7.

where $\theta_1 =$ the angle of incidence and $\theta_2 =$ the angle of refraction. By using Snell's law of refraction (c.f. eq. (1.16)) $n_1 \sin \theta_1 = n_2 \sin \theta_2$, we get the following alternative expressions for the reflection coefficients

$$r_p = \frac{n^2 \cos \theta - \sqrt{n^2 - \sin^2 \theta}}{n^2 \cos \theta + \sqrt{n^2 - \sin^2 \theta}} \tag{9.22a}$$

$$r_n = \frac{\cos \theta - \sqrt{n^2 - \sin^2 \theta}}{\cos \theta + \sqrt{n^2 - \sin^2 \theta}} \tag{9.22b}$$

where $n = n_2/n_1$, is the relative refractive index and where we also have dropped the subscript 1 on the angle of incidence.

The corresponding amplitude transmission coefficients are given as

$$t_p = \frac{2 \sin \theta_2 \cos \theta_1}{\sin(\theta_1 + \theta_2) \cos(\theta_1 - \theta_2)} \tag{9.23a}$$

$$t_n = \frac{2 \sin \theta_2 \cos \theta_1}{\sin(\theta_1 + \theta_2)} \tag{9.23b}$$

with the alternative expressions

$$t_p = \frac{2n \cos \theta}{n^2 \cos \theta + \sqrt{n^2 - \sin^2 \theta}} \tag{9.24a}$$

$$t_n = \frac{2 \cos \theta}{\cos \theta + \sqrt{n^2 - \sin^2 \theta}} \tag{9.24b}$$

Figure 9.8 shows plots of r_p and r_n versus the angle of incidence for $n = 1.52$ (typically for an air–glass interface). We see that $r_p = 0$ at an angle of incidence equal to θ_B. From Equation 9.21(a) we see that this occurs when

$$\theta_1 + \theta_2 = \pi/2 \tag{9.25}$$

which, when inserted into Snell's law gives

$$\tan \theta_B = \frac{n_2}{n_1} \tag{9.26}$$

θ_B is called the polarization angle or the Brewster angle.

This is our reason for wearing polarization sunglasses. When the sunlight, which is unpolarized, strikes a dielectric surface (like the sea) near the Brewster angle, the reflected light becomes linearly polarized. When the transmission axis of the polarization sunglasses is properly oriented, this specularly polarized reflected component will be blocked out.

As mentioned in Section 1.8, when light is incident from a medium of higher onto a medium of lower refractive index, we get total internal reflection at the critical angle given by

$$\sin \theta_c = \frac{n_2}{n_1} \tag{9.27}$$

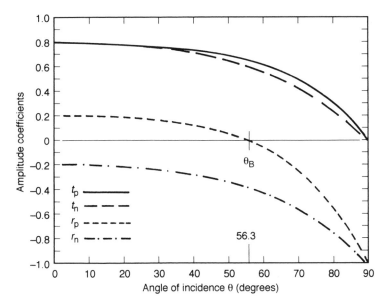

Fig. 9.8. *The amplitude coefficients r_p, r_n, t_p and t_n as a function of incident angle θ at a dielectric interface (air-glass, $n_2/n_1 = 1.5$)*

The field amplitude of a plane wave transmitted into the second medium can be described by (see Equation (1.9))

$$u_t = U_t e^{ik_2(x\sin\theta_2 + z\cos\theta_2)} \tag{9.28}$$

When using Snell's law we get

$$\sin\theta_2 = \frac{n_1}{n_2}\sin\theta_1 = \frac{\sin\theta_1}{n} \tag{9.29a}$$

$$\cos\theta_2 = \pm\sqrt{1 - \left(\frac{\sin\theta_1}{n}\right)^2} \tag{9.29b}$$

or, since we are concerned with the case where $\sin\theta_1 > n$

$$\cos\theta_2 = \pm i\sqrt{\left(\frac{\sin\theta_1}{n}\right)^2 - 1} \equiv \pm i\beta \tag{9.30}$$

Hence

$$u_t = U_t e^{ik_2 x\sin\theta_1/n} e^{\pm k_2\beta z} \tag{9.31}$$

Neglecting the positive exponential which is physically untenable we have a wave whose amplitude drops off exponentially as it penetrates the less dense medium. The field advances in the x-direction as a so-called surface or evanescent wave. Notice that the wavefronts or surfaces of constant phase (parallel to the yz-plane) are perpendicular to the surfaces of constant amplitude (parallel to the xy-plane) and therefore the wave is said to be inhomogeneous. Its amplitude decays rapidly in the z-direction, becoming negligible at a distance into the second medium of only a few wavelengths.

Precautions should be taken when calculating the transmittance of the interface. Transmittance is the ratio of the transmitted over the incident flux and is given by

$$T = \frac{I_t \cos \theta_2}{I_i \cos \theta_1} \tag{9.32}$$

where the projected areas of the incident beams are taken into account since they are unequal. Since u_i and u_t are propagating in media of different refractive indices we must also take into account the correct proportionality factor between the field amplitude squared and the irradiance, see Section 1.7, Equation (1.15). Then we get

$$\frac{I_t}{I_i} = \frac{n_2 |U_t|^2}{n_1 |U_i|^2} \tag{9.33}$$

and the transmittance

$$T = \frac{n_2 \cos \theta_2 |U_t|^2}{n_1 \cos \theta_1 |U_i|^2} = \frac{n_2 \cos \theta_2}{n_1 \cos \theta_1} t^2 \tag{9.34}$$

Since both the incident and reflected beams are propagating in the same medium and their projected areas are equal, we get for the reflectance

$$R = r^2 \tag{9.35}$$

or in component forms

$$R_p = r_p^2 \tag{9.36a}$$

$$R_n = r_n^2 \tag{9.36b}$$

$$T_p = \frac{n_2 \cos \theta_2}{n_1 \cos \theta_1} t_p^2 = \frac{\sqrt{n^2 - \sin^2 \theta}}{\cos \theta} t_p^2 \tag{9.36c}$$

$$T_n = \frac{n_2 \cos \theta_2}{n_1 \cos \theta_1} t_n^2 = \frac{\sqrt{n^2 - \sin^2 \theta}}{\cos \theta} t_n^2 \tag{9.36d}$$

It can be shown that

$$R_p + T_p = 1 \tag{9.37a}$$

$$R_n + T_n = 1 \tag{9.37b}$$

as it should be.

9.5   THE JONES MATRIX FORMALISM OF POLARIZED LIGHT

Equation (9.4) can be written in a more compact form as

$$|U\rangle = e^{i\beta/2} \begin{pmatrix} u_x \\ u_y \end{pmatrix} \tag{9.38}$$

where

$$u_x = U_x e^{-i\delta/2} \tag{9.39a}$$

$$u_y = U_y e^{i\delta/2} \tag{9.39b}$$

$|U\rangle$ is called a Jones vector or state vector, representing the state of polarization of the wave. Evidently, the state of polarization, and thereby the Jones vector, remain unchanged by a multiplicative constant. Therefore

$$|U'\rangle = c|U\rangle \tag{9.40}$$

and $|U\rangle$ are equal state vectors representing the same state of polarization. Equation (9.38) can be decomposed to read

$$|U\rangle = U_x e^{-i\delta/2}|P_x\rangle + U_y e^{i\delta/2}|P_y\rangle \tag{9.41}$$

where

$$|P_x\rangle = \begin{pmatrix} 1 \\ 0 \end{pmatrix}. \tag{9.42a}$$

$$|P_y\rangle = \begin{pmatrix} 0 \\ 1 \end{pmatrix} \tag{9.42b}$$

are base vectors representing waves, linearly polarized in the x- and y-directions respectively. Here we have omitted the common phase factor $e^{i\beta/2}$ for reasons given after Equation (9.5). Equation (9.41) is a general expression for an arbitrary state of polarization decomposed into the orthogonal basis $|P_x\rangle, |P_y\rangle$.

As discussed in Section 9.1, Equation (9.41) represents a wave, linearly polarized at an angle ϕ to the x-axis, and is termed a P-state, when $\delta = 0$ and $U_x = \cos\phi$, $U_y = \sin\phi$, which yields

$$|P\rangle = \cos\phi|P_x\rangle + \sin\phi|P_y\rangle \tag{9.43}$$

For circularly polarized states, we further have $\delta = \pm\pi/2$ and $U_x = U_y = 1/\sqrt{2}$ (normalized intensity). Specifically, a right circularly polarized wave (an R-state) is given by

$$|R\rangle = \frac{1}{\sqrt{2}}(e^{-i\pi/4}|P_x\rangle + e^{i\pi/4}|P_y\rangle) \tag{9.44a}$$

and a left circularly polarized wave (an L-state) by

$$|L\rangle = \frac{1}{\sqrt{2}}(e^{i\pi/4}|P_x\rangle + e^{-i\pi/4}|P_y\rangle) \tag{9.44b}$$

A state vector may equally well be represented by a set of orthonormal base vectors other than $|P_x\rangle, |P_y\rangle$. For example $|R\rangle$ and $|L\rangle$ will form such a set. That two vectors

$$|A\rangle = \begin{pmatrix} a_1 \\ a_2 \end{pmatrix} \tag{9.45a}$$

$$|B\rangle = \begin{pmatrix} b_1 \\ b_2 \end{pmatrix} \tag{9.45b}$$

are orthonormal means that the scalar products

$$\langle A|A\rangle = \langle B|B\rangle = 1 \qquad (9.46a)$$

$$\langle A|B\rangle = \langle B|A\rangle = 0 \qquad (9.46b)$$

Here $\langle A|$ and $\langle B|$ means the row matrices

$$\langle A| = (a^* a_2^*) \qquad (9.47a)$$

$$\langle B| = (b_1^* b_2^*) \qquad (9.47b)$$

and the scalar product is given by ordinary matrix multiplication, e.g.

$$\langle A|B\rangle = (a_1^* a_2^*)\begin{pmatrix} b_1 \\ b_2 \end{pmatrix} = a_1^* b_1 + a_2^* b_2 \qquad (9.48)$$

where the asterisks denote complex conjugation, see Appendix A.

We have seen that, for example, for a P_x-state, there exists a P_x-filter, i.e. a linear polarizer with its transmission direction along the x-axis, for an R-state there exists an R-filter, i.e. a right circular polarizer, and so on. Generally, it can be shown that to elliptically polarized light, an E-state, it is possible to construct a corresponding E-filter. The physical interpretation of the scalar product $\langle A|B\rangle$ is therefore given as the probability that an A-filter will be transmitted by a B-state and the transmitted intensity is given as the absolute square of the probability, i.e.

$$I = |\langle A|B\rangle|^2 \qquad (9.49)$$

As an example, we calculate the intensity when a P_x-filter is transmitted by a P-state given by Equation 9.43

$$I = |\langle P_x|P\rangle|^2 = |\cos\phi\langle P_x|P_x\rangle + \sin\phi\langle P_x|P_y\rangle|^2 = \cos^2\phi \qquad (9.50)$$

since

$$\langle P_x|P_x\rangle = 1 \quad \text{and} \quad \langle P_x|P_y\rangle = 0$$

which again gives Malus' law.

When the state of polarization of a light wave is changed by a polarization filter (or any other optical phenomenon) this change can be described as an input-output relation by a 2×2 Jones matrix, viz.

$$|E_u\rangle = M|E_i\rangle \qquad (9.51)$$

where $|E_i\rangle$ and $|E_u\rangle$ are the state vectors of the incoming and outcoming light respectvely, and M is the matrix representing the polarization change. It is the power of Jones matrix theory that when light passes through several polarizing filters in succession, represented by, for example, the matrices $M_1, M_2, \ldots, M_n$, the total polarizing effect is described by the product of the matrices of the individual filters, i.e.

$$|E_u\rangle = M_n \ldots M_2 M_1|E_i\rangle \qquad (9.52)$$

Note the order of matrices of the product.

It is easily verified that the Jones matrix for a linear polarizer with its transmission direction along the x-axis represented in the $|P_x\rangle$, $|P_y\rangle$ basis, must be

$$P_x = \begin{pmatrix} 1 & 0 \\ 0 & 0 \end{pmatrix} \qquad (9.53)$$

and that of a retarder with retardance δ with its fast axis along the x-axis must be

$$M_x = \begin{pmatrix} e^{-i\delta/2} & 0 \\ 0 & e^{i\delta/2} \end{pmatrix} \qquad (9.54)$$

To find the general expressions of these matrices when their axes are oriented at an arbitrary angle ϕ to the x-axis, we proceed as follows. First rotate the coordinate system an angle ϕ such that the x'-axis coincides with the filter axis, then apply the matrix of the filter and finally rotate the coordinate system back to its original position. This is expressed mathematically as:

For the polarizer

$$P = R^{-1}P_x R \qquad (9.55)$$

For the retarder

$$M = R^{-1}M_x R \qquad (9.56)$$

where

$$R = \begin{pmatrix} \cos\phi & \sin\phi \\ -\sin\phi & \cos\phi \end{pmatrix} \qquad (9.57)$$

is the rotation matrix and

$$R^{-1} = \begin{pmatrix} \cos\phi & -\sin\phi \\ \sin\phi & \cos\phi \end{pmatrix} \qquad (9.58)$$

is its inverse. This gives

$$P = \begin{pmatrix} a & b \\ b & c \end{pmatrix} \qquad (9.59)$$

where

$$a = \cos^2\phi$$
$$b = \sin\phi\cos\phi$$
$$c = \sin^2\phi \qquad (9.60)$$

and

$$M = \begin{pmatrix} p & q \\ q & r \end{pmatrix} \qquad (9.61)$$

where

$$p = e^{-i\delta/2}\cos^2\phi + e^{i\delta/2}\sin^2\phi$$
$$q = -2i\sin(\delta/2)\sin\phi\cos\phi \qquad (9.62)$$
$$r = e^{-i\delta/2}\sin^2\phi + e^{i\delta/2}\cos^2\phi$$

As examples, by putting $\delta = \pi/2$ and $\delta = \pi$ we find the matrices for the $\lambda/4$ and $\lambda/2$ plates respectively.

9.6   PHOTOELASTICITY

9.6.1   Introducton

Photoelasticity is a method for stress analysis of specimens subject to load. The standard methods rely on the technique of making a model which is a copy (often on a reduced scale) of the specimen under investigation.

This model, usually made from plastics such as epoxy or polyester resins, becomes birefringent when subject to loads. Thus, a two-dimensional photoelastic model exerted by forces in its own plane will behave as a general retarder, the retardance and the direction of the retarder axes being continuous variables across the model plane. The retardance δ is given as

$$\delta = kC(\sigma_1 - \sigma_2)t \tag{9.63}$$

where σ_1 and σ_2 are the principal stresses, C is the stress-optic coefficient characteristic of the model material and t is the thickness of the model. The retarder axes h_1, h_2 coincide with the axes of the principal stresses σ_1, σ_2.

When this loaded model is placed in a polariscope which consists of a light source and properly arranged polarizing elements, a system of fringes is observed. In the next subsections we shall find the relation between this fringe system and the stresses.

In a related technique termed the photoelastic coating method, a birefringent coating is attached to the real specimen. The optical principles here are essentially the same as for the standard methods except that the fringes have to be observed in reflection instead of transmission.

9.6.2   The Plane Polariscope

In Figure 9.9 the photoelastic model is placed in plane 2 between two linear polarizers in planes 1 and 3. This configuration is termed a plane polariscope. When the two polarizers are crossed (i.e. their transmission axes are inclined at 90°) as in the figure, we have a dark-field plane polariscope, a term which refers to the field without the model. When the two polarizers are parallel, we have a light-field plane polariscope.

Let us for the time being assume the model to be a uniform retarder with its principal axis h_1 making an angle α to the x-axis and try to find the intensity of the light behind plane 3.

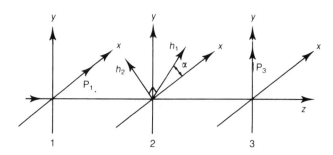

Fig. 9.9.   *Plane polariscope. Schematic*

The transmission direction of P_1 is along the x-axis. Hence the field behind P_1 is

$$\mathbf{u}_1 = \mathbf{e}_x \tag{9.64}$$

where we have put the amplitude equal to 1. The components of u_1 along h_1 and h_2 are $\cos \alpha$ and $-\sin \alpha$ respectively. The field behind the retarder therefore becomes

$$\mathbf{u}_2 = \cos \alpha e^{-i\delta/2} \mathbf{e}_1 - \sin \alpha e^{i\delta/2} \mathbf{e}_2 \tag{9.65}$$

where $\mathbf{e}_1$ and $\mathbf{e}_2$ are the unit vectors along h_1 and h_2. The transmission direction of P_3 is along the y-axis. To find the field behind plane 3, we therefore have to calculate the component of $\mathbf{u}_2$ along the y-axis which gives

$$\mathbf{u}_3 = \sin \alpha \cos \alpha (e^{-i\delta/2} - e^{i\delta/2}) \mathbf{e}_y \tag{9.66}$$

and hence the intensity

$$I_3 = |\mathbf{u}_3|^2 = 4 \sin^2 \alpha \cos^2 \alpha \sin^2 (\delta/2)$$
$$= \sin^2 2\alpha \sin^2 (\delta/2) \tag{9.67}$$

It is easily verified that if the transmission direction of P_3 is along the x-axis, the intensity becomes

$$I'_3 = (1 - \sin^2 2\alpha \sin^2 (\delta/2)) \tag{9.68}$$

These calculations can be written in a very compact and elegant form using the state vectors $|P_x\rangle$ and $|P_y\rangle$ and the matrix M of a retarder (Equation (9.61); cf. Equation (9.49))

$$I_3 = |\langle P_y | M | P_x \rangle|^2 = |q|^2 = \sin^2(2\alpha) \sin^2(\delta/2) \tag{9.69}$$

$$I'_3 = |\langle P_x | M | P_x \rangle|^2 = |p|^2 = 1 - \sin^2(2\alpha) \sin^2(\delta/2) \tag{9.70}$$

For a uniform retarder the intensity across the field of view is uniform and determined by Equations (9.69) and (9.70). For a photoelastic model, however, both α and δ vary over the xy-plane resulting in a system of fringes across the field of view. From Equation (9.69) and (9.70) we see that this fringe system consists of two sets: one depending on α, the other on δ. In a dark-field polariscope we see from Equation (9.69) that $I_3 = 0$ when $\alpha = 0$, i.e. we get a dark fringe. These fringes depending on α are called isoclines, which means 'equal inclinations', and are loci of points where the principal stress axes coincide with the axes of the polarizers. They show up as broad dark bands on the model. By synchronous rotation of the two polarizers, we vary α and the isoclinics move, thereby determining the direction of the principal stresses over the whole model. From Equation (9.69) and (9.70) we see that $I_3 + I'_3 = 1$. A dark isocline in a dark-field polariscope therefore turns into a bright isocline in a light-field polariscope. The other fringe system depending on δ does not change its position by varying α. We will examine these fringes more closely in the next subsection.

9.6.3 The circular polariscope

Figure 9.10 shows the same set-up as in Figure 9.9 except that two quarterwave plates with their axes oriented $45°$ to the transmission directions of the polarizers are placed between the polarizers and the retarder. We thus have a circular polarizer on each side of

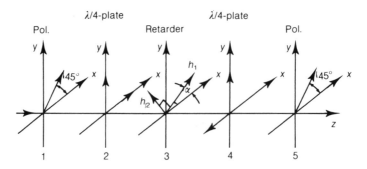

Fig. 9.10. *Circular polariscope. Schematic*

the retarder (the model). This configuration is called a circular polariscope, and is termed a dark-field polariscope when the two circular polarizers have opposite handedness and a light-field polariscope when the two circular polarizers have the same handedness.

To calculate the intensity behind plane 5, we now may derive the field amplitude behind each element in steps, in the same manner as above. This, however, is a tedious task. A much simpler way is to apply Jones matrix algebra, which gives:

For circular polarizers of opposite handedness

$$I_s = |\langle L|M|R \rangle|^2 = |p - r - 2iq|^2 = \sin^2(\delta/2) = \tfrac{1}{2} - \tfrac{1}{2}\cos\delta \qquad (9.71)$$

For circular polarizers of the same handedness

$$I'_5 = |\langle R|M|R \rangle|^2 = |p + r|^2 = \cos^2(\delta/2) = \tfrac{1}{2} + \tfrac{1}{2}\cos\delta \qquad (9.72)$$

From these equations we see that in a circular polariscope the fringe pattern is only dependent on δ, i.e. the isoclinics vanish. From Equation (9.63) we further see that δ and thereby the intensity is dependent on the wavelength λ. Therefore, by using a white light source in a circular polariscope, the resulting pattern will show up as fringes of different colours. These fringes, therefore are termed isochromatics which means equal colours. As can be seen from Equation (9.63), these isochromatics are loci of equal principal stress difference $\sigma_1 - \sigma_2$.

From Equation (9.71) we see that in a dark-field circular polariscope, we get a dark isochromatics when

$$\delta = \frac{2\pi}{\lambda} C(\sigma_1 - \sigma_2)t = 2\pi \cdot n, \qquad n = 0, 1, 2, \ldots \qquad (9.73)$$

i.e. when

$$\sigma_1 - \sigma_2 = n \cdot S \qquad (9.74)$$

and bright isochromatics when

$$\frac{2\pi}{\lambda} C(\sigma_1 - \sigma_2)t = 2\pi(n + \tfrac{1}{2}), \qquad n = 0, 1, 2, \ldots \qquad (9.75)$$

i.e. when

$$\sigma_1 - \sigma_2 = (n + \tfrac{1}{2})S \qquad (9.76)$$

where n is the order of the isochromatics and where we have put

$$S = \frac{\lambda}{Ct} \tag{9.77}$$

When using the same material of the same thickness and a monochromatic light source of the same wavelength, S is a system constant. It is usually calibrated by measuring the fringe order in the centre of a diametrally loaded circular disc and inserting its value into the formula

$$S = \frac{8F}{n\pi D} \tag{9.78}$$

where F is the applied force, D is the diameter of the disc, and n is the measured fringe order.

The principal stress difference in an arbitrary model is then found by just counting the fringe order n using eqs. (9.74)–(9.76). To do this, of course, one needs to know where to start the counting, i.e. which is the zero-order fringe. From eq. (9.73) we see that for $n = 0$, δ becomes independent of the wavelength. Therefore, by using a white light source, fringes of order $n > 0$ become coloured while the zero-order fringe is easily recognized as a dark colourless fringe. The higher-order fringes are most accurately determined by using monochromatic light, often a sodium lamp. A standard polariscope is therefore usually equipped with two light sources.

Figure 9.11 shows a typical pattern of isochromatics in a dark-field polariscope. The model is a strip with a hole under uniaxial tension.

Finally, we mention that on a load-free boundary of the model, the principal stress normal to the boundary is zero, and the value of the other principal stress is found directly.

9.6.4 Detection of isochromatics of fractional order. Compensation

We have shown that in a circular polariscope, an isochromatic fringe pattern giving the stress distribution is observed. It may happen, however, that an interesting measuring point lies between two isochromatic fringes. To determine the fringe order with sufficient accuracy at such a point there exist several methods. These are as follows.

(1) Varying the load (whose value must be known) until an isochromatic fringe intersects the point of interest. This is not always possible.

(2) Colour matching, i.e. knowing the change in colour with increasing order using a white light source or a spectral lamp. This method relies on human judgement of colours and is possible only for low orders.

(3) Photometric methods, i.e. accurately measuring the intensity at the poin of interest when rotating the polarizing elements with known amounts. These methods have some inherent sources of error.

(4) Compensation. This can be done by first orienting, for example, a Babinet–Soleil compensator parallel to the direction of the principal stresses, i.e. the isoclinics, and then adjust the compensator's retardance until the measurement point becomes dark. Non-expensive compensators can be made from loaded strips of photoelastic material.

Fig. 9.11. *Dark-field isochromatic patern from a strip with a hole under uniaxial tension*

The most common method of compensation, however, is by means of quarterwave plates, known as the Tardy of Senarmont methods. Here we shall describe the latter in more detail.

In Figure 9.12(a) the point of interest P lies between two isochromatics of order n and $n + 1$. To determine the order ($= n + p$ where $0 < p < 1$) at point P we proceed as follows:

(1) Change to a dark-field plane polariscope configuration (Figure 9.9) and rotate the polarizers P_1 and P_3 in synchronization until a bright isoclinic intersects P (see Figure 9.12(b)). We then have $\alpha = 45°$ and $I_3 = \sin^2(\delta/2)$ (cf. Equation 9.69)).

(2) Place a quarterwave plate with its axes parallel to the transmission directions of P_1 and P_3 between the model and P_3, i.e. in a plane 2b between planes 2 and 3. To find the field

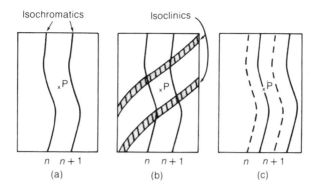

Fig. 9.12

amplitude $\mathbf{u}_{2b}$ behind the quarterwave plate, we consider Equation 9.65 which gives the field incident on the quarterwave plate. We find (for $\alpha = 45°$)

$$\mathbf{u}_{2b} = \cos \delta e^{i\pi/4}\mathbf{e}_x - i \sin \delta e^{-i\pi/4}\mathbf{e}_y \qquad (9.79)$$

(3) Rotate P_3 (alone) an angle χ. The field amplitude u_3 behind P_3 then becomes

$$\mathbf{u}_3 = -(\cos \delta \sin \chi e^{i\pi/4} + i \sin \delta \cos \chi e^{-i\pi/4})\mathbf{e}_x$$
$$= -\sin(\chi + \delta)e^{i\pi/4}\mathbf{e}_x \qquad (9.80)$$

where $\mathbf{e}_x$ is he unit vector along the transmission direction of P_3. The intensity now becomes equal to

$$I_3 = |\mathbf{u}_3|^2 = \sin^2(\chi + \delta) \qquad (9.81)$$

For the nth-order isochromatics, $\delta = n \cdot 2\pi$ and

$$I_3 = \sin^2 \chi \qquad (9.82)$$

which for $\chi = 0$ gives $I_3 = 0$ as it should. By rotating P_3 an angle $\chi = \pi$, we again have $I_3 = 0$. We will observe that during this rotation, the nth-order isochromatics would have moved to the initial position of the $(n + 1)$th-order isochromatics.

For an intermediate value χ_p of the rotation angle, the nth-order isochromatics will intersect our point P (see Figure 9.12(c). A rotation angle $\chi = \pi$ therefore corresponds to increasing the order number by unity, while a rotation angle χ_p corresponds to an increase of the order number equal to

$$p = \chi_p/\pi \qquad (9.83)$$

where χ_p is measured in radians. In conclusion, we have found the isochromatic order in point P to be $n + \chi_p/\pi$.

In practice, this procedure can be executed as follows. The polarizer P_1, the model, the quarterwave plate Q and the analyser P_3 are place in that order as described above.

P_1 and P_2 are crossed (dark-field polariscope) and the axes of Q are parallel to the transmission directions of P_1 and P_3. In this position, the quarterwave plate does not affect the intensity (see Equation (9.17)). The scales of P_1, Q and P_3 are divided to read 1 for a 180° rotation.

Initially, all scales are set to zero. The filters are coupled together, e.g. by means of toothwheel transmissions. Firstly, all three filters are rotated until a dark isoclinic intersects P and the resulting scale reading is a. This is taken as the first step because it is easier to determine the intersection of a dark isoclinic rather than a bright one. Then the filters are rotated 45°, i.e. to a scale reading equal to $b = a + 0.25$. We then have reached point (1) in the description above, i.e. a bright isoclinic intersects P. Finally, the analyser P_3 is released and rotated alone (point 3) until the nth-order isochromatic intersects P and the new scale reading is c. The isochromatic order at P is then given as

$$n + |c - b| \tag{9.84}$$

9.7 HOLOGRAPHIC PHOTOELASTICITY

As shown in the previous chapter, the directions and the difference $\sigma_1 - \sigma_2$ of the principal stresses can be determined by using a standdard polariscope. For a complete solution of the stress distribution inside a two-dimensional model, however, the absolute values of σ_1 and σ_2 must be known.

Up to now we have omitted a common phase factor $e^{i\beta/2}$ imposed on the light when passing through the stressed model. This phase is unaffected by the birefringence of the model, and is given by

$$\beta = (2\pi/\lambda)[K(\sigma_1 + \sigma_2) + 2n_0]t \tag{9.85}$$

where K is a constant of the model material and n_0 is the refractive index of the unstressed model. If β could be determined, the principal stress sum $\sigma_1 + \sigma_2$ is given, which together with the values of $\sigma_1 - \sigma_2$ gives a complete solution of the stress-distribution problem. This phase can be found by interferometric methods. Here the method of holographic interferometry will be considered in more detail.

For this purpose, a set-up like that shown in Figure 9.13 can be applied. Here the light passing through the model placed in a light-field circular polariscope constitutes the

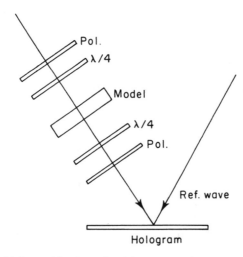

Fig. 9.13. *Holographic photoelesticity set up. λ/4 = quarterwave plates*

object wave. The method consists of making two exposures of the model, first in its unloaded and then in its loaded condition. To get a registration of the unloaded model, it is therefore essential to have a light-field polariscope.

The matrix representing the unloaded model will be

$$M_o = e^{i\phi_o} I \tag{9.86}$$

where $\phi_0 = kn_0 t$ is the uniform phase due to the unloaded model and I is the identity matrix

$$I = \begin{pmatrix} 1 & 0 \\ 0 & 1 \end{pmatrix} \tag{9.87}$$

The complex amplitude recorded in the first hologram exposure therefore can be written as the state vector

$$|S_0\rangle = SM_0|R\rangle = e^{i\phi_o}|P_x\rangle \tag{9.88}$$

where we have assumed the incident light to be in a right circular state and where S represents the quarterwave plate-polarizer combination behind the model.

The field amplitude when the model is loaded in the second exposure then becomes

$$|S\rangle = SM|R\rangle = e^{i\beta/2} \cos(\delta/2)|P_x\rangle \tag{9.89}$$

With a reference wave polarized in a P_x-state, these two exposures will be recorded with maximum modulation. The second polarizer in Figure 9.13 can be removed since the reference wave will act as an analyser when interfering with the object wave. By preparing the reference wave in a right circular polarized state, the second quarterwave plate can also be removed. But then the recorded modulation will be reduced and unwanted polarization effects will occur.

By reconstruction of this doubly exposed hologram, the reconstructed field will be proportional to the sum of the partial exposures and the intensity becomes

$$\begin{aligned} I &= |e^{i\phi_o} + e^{i\beta/2} \cos(\delta/2)|^2 \\ &= 1 + 2\cos(\delta/2)\cos(\phi_0 - \beta/2) + \cos^2(\delta/2) \\ &= 1 + 2\cos\left[k\frac{C}{2}(\sigma_1 - \sigma_2)t\right]\cos\left[k\frac{K}{2}(\sigma_1 + \sigma_2)t\right] + \cos^2\left[k\frac{C}{2}(\sigma_1 - \sigma_2)t\right] \end{aligned} \tag{9.90}$$

The resulting fringe pattern therefore consists of two types: the isochromatics depending on $\sigma_1 - \sigma_2$ and the so-called isopachics depending on $\sigma_1 + \sigma_2$.

From the second term of Equation 9.90, we see that the isopachics are modulated by the isochromatics and the analysis of this combined pattern might in some cases become quite complicated.

A dark isochromatic occurs whenever

$$\delta = (2n+1)\pi, \quad \text{for} \quad n = 0, 1, 2, \dots \tag{9.91}$$

with a resulting intensity

$$I = 1 \tag{9.92}$$

The dark isochromatics therefore possess non-zero intensity and show up as half-tone fringes in the resulting pattern. A bright isochromatic occurs whenever

$$\delta = n \cdot 2\pi, \qquad \text{for} \quad n = 0, 1, 2, \dots \tag{9.93}$$

with a resulting intensity

$$I = 2[1 + (-1)^n \cos(\phi_0 - \beta/2)] \tag{9.94}$$

Assume for a moment n to be an even number k. The intensity is then

$$I_k = 2[1 + \cos(\phi_0 - \beta/2)] \tag{9.95}$$

To simultaneously have a drak isopachic we must demand

$$\phi_0 - \beta/2 = (m + \tfrac{1}{2})2\pi, \qquad \text{for} \quad m = 0, 1, 2, \dots \tag{9.96}$$

which gives $I_{k,m} = 0$. When tracing this dark isopachic from the kth bright isochromatic into the next $(k+1)$th bright isochromatic, the intensity becomes $I_{k+1,m} = 4$ which implies that the isopachic has changed from a dark to a bright fringe in crossing the isochromatic. Moreover, the condition of a dark isopachic now becomes

$$\phi_0 - \beta/2 = m \cdot 2\pi \tag{9.97}$$

which shows that the isopachic has changed by one half-order in crossing the dark isochromatic.

These characteristics of the combined pattern are shown in Figure 9.14 for the case where the isochromatics and the isopachics cross each other nearly perpendicularly.

Fig. 9.14

Fig. 9.15. *Combined isochromatics–isopachics patttern obtained with double-exposure holography*

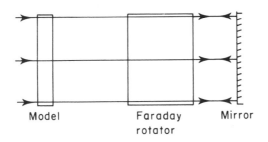

Fig. 9.16

Examples of this phase shift are also seen on the photograph in Fig. 9.15. In the case of perpendicular crossing of the two patterns, their analysis becomes fairly easy, but when the two patterns are nearly parallel, the quantitative interpretation can give erroneous results. Attempts have therefore been made to separate the two patterns. One solution is to let the object wave pass twice through the model and a Faraday rotator as shown Figure 9.16. By passing the rotator the orthogonal polarizations are reversed and the birefringence effect is cancelled in the second pass of the model resulting in a isopachics pattern only.

9.8 THREE-DIMENSIONAL PHOTOELASTICITY

9.8.1 Introduction

Up to now we have considered two-dimensional states of stress, i.e. the stress in the z-direction is constant. When the stress also varies in this direction, we have a three-dimensional stress state. This cannot generally be investigated by the two-dimensional procedure of observing the isochromatics and isoclinics in a polariscope. This is because the birefringence effect integrated over the optical path is generally so complex that it is impossible to relate it to the stresses which produce it.

Several methods are available for the investigation of three-dimensional problems. Below we briefly discuss two of them, the frozen stress method and the scattered light method.

9.8.2 The frozen stress method

This is the most widely used three-dimensional method; it is restricted, however, to static loading by external forces or constant body forces such as gravitational or centrifugal forces. In this method, advantage is taken of the multiphase (diphase) nature of plastics used as model materials to conserve strain and birefringence in the model after the load has been removed. Then, slices can be cut from the model and analysed in a polariscope as a two-dimensional model without disturbing the frozen-in stress pattern.

The procedure of strain freezing consists of heating the model to a temperature slightly above the softening point or critical temperature followed by slow cooling under load to room temperature. In practice, it is most convenient to load the model before placing it in the oven. While the rate of heating is immaterial, the cooling must be carried out slowly, typically at less than 2°C per hour.

In the slicing process, sharp tools should be used to avoid overheating, which may disturb the frozen stress distribution.

Consider a slice cut from a three-dimensional model in such a way that the directions of two of the three principal stresses, e.g. σ_1 and σ_2, lie in the plane of the slice (see Figure 9.17). This pre-supposes that the direction of at least one of the principal stresses, e.g. σ_3, is known, which is frequently the case, for example on a section of symmetry or

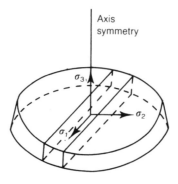

Fig. 9.17

a free surface of the model. By placing this slice in a polariscope, the isochromatics will determine $\sigma_1 - \sigma_2$ while the isoclinics will indicate their directions. To determine the principal stress difference in the other two principal planes, a subslice may be cut from the original slice parallel to one of the principal stresses, say σ_1 (see Figure 9.17). If this subslice is observed in a polariscope in the direction of σ_2, the resulting isochromatics will give the value of $\sigma_1 - \sigma_3$. The remaining value $\sigma_2 - \sigma_3$ may then be found by subtracting the second value from the first.

When none of the directions of the three principal stresses is *a priori* known, the situation is more complicated. For the detailed description of the various methods for solving this problem, we direct the reader to the specialized literature given in the Bibliography.

9.8.3   The scattered light method

If a beam of light passes through a transparent isotropic medium, it will be scattered by small particles in suspension or by the molecules of the medium. The field amplitude of the light scattered from a certain point, will be proportional to the component of the amplitude of the incident light normal to the scattering direction.

Consider Figure 9.18 where a thin pencil of light propagates in the z-direction through a three-dimensional photoelastic model. At each point along this primary beam we observe the scattered light in a direction normal to z, say the x-axis. The field amplitude of the scattered light then becomes equal to the y-component of the field of the primary beam at the scattering point. Now assume that the principal stresses, say σ_1 and σ_2, are normal to the z-axis, and that their directions remain constant along the primary beam and make

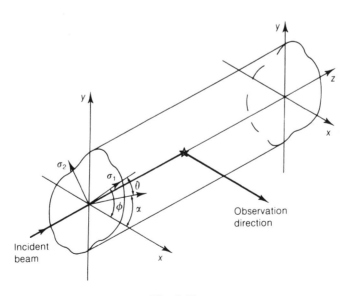

Fig. 9.18

an angle ϕ to the x-axis. Further assume that the incident light is linearly polarized at an angle α to the x-axis. The phase difference between the field components parallel to σ_1 and σ_2 accumulated from the point of entry to the scattering point we denote by Δ. The calculation of the scattered intensity now becomes equivalent to the problem of calculating the intensity of the light transmitting a retarder given by Equation (9.61) placed between a polarizer

$$|P\rangle = \begin{pmatrix} \cos \alpha \\ \sin \alpha \end{pmatrix}$$

(9.98)

and an analyser $|P_y\rangle = \begin{pmatrix} 0 \\ 1 \end{pmatrix}$ (see Figure 9.18). This gives

$$I = |\langle P_y|M|P\rangle|^2 = |q \cos \alpha + r \sin \alpha|^2$$
$$= \sin^2 \alpha + \sin^2(\Delta/2) \sin 2\phi \sin 2(\phi - \alpha)$$

(9.99)

which, by introducing the angle $\theta = \phi - \alpha$ between the retarder axis and the polarization direction, becomes

$$I = \tfrac{1}{2}(1 - \cos 2\phi \cos 2\theta - \sin 2\phi \sin 2\theta \cos \Delta)$$

(9.100)

Therefore, by observing the scattered light in a direction perpendicular to the primary beam along its path, the intensity will vary from

$$I_{\min} = \tfrac{1}{2}[1 - \cos 2(\phi - \theta)] = \sin^2 \alpha$$

(9.101)

at points where

$$\Delta = n \cdot 2\pi \quad \text{for} \quad n = 0, 1, 2, \ldots$$

to

$$I_{\max} = \tfrac{1}{2}[1 - \cos 2(\phi + \theta)]$$

(9.102)

at points where

$$\Delta = (2n + 1)\pi$$

This gives for the visibility

$$V = \frac{I_{\max} - I_{\min}}{I_{\max} + I_{\min}} = \frac{\sin 2\phi \sin 2\theta}{1 - \cos 2\phi \cos 2\theta}$$

(9.103)

From this we see that the fringe visibility is maximum (i.e. unity) when $\theta = \phi$, i.e. when the direction of observation is parallel to the polarization direction or when the axis of σ_1 bisects the angle between the polarization and observation directions. In the first case, the intensity becomes

$$I = \tfrac{1}{2} \sin^2 2\phi (1 - \cos \Delta)$$

(9.104)

and in the second case ($\theta = -\phi$)

$$I = \tfrac{1}{2} \sin^2 2\phi (1 + \cos \Delta)$$

(9.105)

Moreover, the maximum variation of I occurs for $\sin 2\phi = 1$, i.e. when the principal stress axis is inclined $45°$ to the direction of observation.

From Equation (9.100) we see that the intensity is constant and independent of Δ when either the polarization or observation direction is parallel to one of the principal stresses. No interference effects will then be observed along the path of the primary beam. Moreover, if $\theta = \phi = 0°$ or $90°$, the intensity becomes a minimum. The directions of the principal stresses can therefore be determined by making the polarization and observation directions parallel and then rotating the model about the axis of the primary beam until a uniform minimum intensity is observed. With an elemetary path length dz along the primary beam the principal stress difference $\sigma_2 - \sigma_2$ can be regarded as constant. The stress optic law, Equation (9.63) then gives

$$d\Delta = kC(\sigma_1 - \sigma_2)\,dz \tag{9.106}$$

or

$$\sigma_1 - \sigma_2 = \frac{1}{kC}\frac{d\Delta}{dz}$$

Expressed in terms of the fringe order, this equation can be written as

$$\sigma_1 - \sigma_2 = f(dn/dz) \tag{9.107}$$

The principal stress difference at any point along the primary beam is therefore proportional to the gradient or inversely proportional to the spacing of the fringes in the scattered light. The possibility of errors in fringe counting owing to a change of sign of the principal stress difference at any point may be avoided by introducing an additional phase difference by means of a compensator. The fringes on opposite sides of such a point will then move in opposite directions. Alternatively, a change of sign may be indicated by reversal of the colour sequence of the fringes when using white light.

Another variant of the scattered light method is to apply an unpolarized primary beam propagating in, for example, the z-direction as in Figure 9.19. The light scattered in the x-direction from any point will then be linearly polarized along the y-axis. Each point thus acts as a secondary source of linearly polarized light, and if the model is stressed, this light will be influenced by the stresses between the scattering point and the point of exit from the model. If this scattered light emerging from the model is viewed through an analyser, a fringe pattern is seen which has exactly the same meaning as that which would be produced in a plane polariscope. The problem is that the total birefringence will be accumulated from the scattering point to the point of exit from the model.

To apply the method, the primary beam is moved parallel to itself in small steps along the line of observation from the surface towards the interior of the model. For each step, the phase difference of the emergent light components is measured. If the results are plotted against the position of the scattering point, the slope of the resulting curve multiplied by the stress-optic coefficient gives the principal stress difference at that point.

The basic optical requirement in a scattered light polariscope is a fine pencil or thin ribbon of highly collimated light. Since the efficiency of the scattering process is low, a source of high intensity is required. These requirements are met by mercury arc lamps while lasers are particularly suitable. The model is usually placed in an immersion tank to avoid refraction and polarization changes at the model surface.

Provision for rotational and translational movement of the model in the tank is needed. This makes the method rather complicated in practice.

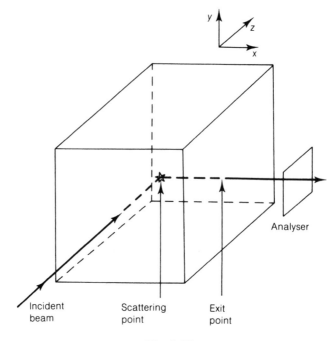

Fig. 9.19

PROBLEMS

9.1 Determine the emergent state in each of the following cases:

(a) A P-state incident on a quarter-wave plate with the line of the P-state midway between the principal axes of the plate.

(b) A P-state incident on a half-wave plate with the line of the P-state midway between the principal axes of the plate.

(c) R- and L-states incident on a quarter-wave plate.

(d) R- and L-states incident on a half-wave plate.

9.2 Given that a general polarization state is given by

$$|E\rangle = A_x e^{-i\phi/2}|P_x\rangle + A_y e^{i\phi/2}|P_y\rangle$$

show that the orthogonal state $|E'\rangle$, such that $\langle E'|E\rangle = 0$ is given by

$$|E'\rangle = A_y e^{-i\phi/2}|P_x\rangle - A_x e^{i\phi/2}|P_y\rangle$$

In the case $|E> = |R\rangle, |E'\rangle = |L\rangle$ show that the above forms are consistent with those given by Equations (9.44a,b).

9.3 If T is the matrix of a polarization element in the $|P_x\rangle, |P_y\rangle$-basis, the matrix T' of the same element in the $|R\rangle, |L\rangle$-basis is given by

$$T' = UTU^+$$

where

$$U = \begin{pmatrix} \langle R|P_x \rangle & \langle R|P_y \rangle \\ \langle L|P_x \rangle & \langle L|P_y \rangle \end{pmatrix}$$

(U^+, the adjoint of U means the transposed and complex conjugate of U)

(a) Calculate U.
(b) Find the expression for the matrix M_x of a retarder given in Equation (9.54) in the $|R\rangle$, $|L\rangle$-basis.

9.4 Just as (linear) retarders resolve beams into P-states and transmit them at different speeds there exist circular retarders which resolve beams into R- and L-states and transmit them at different speeds. Such materials are said to be optically active. The matrix N of a circular retarder with retardance β is therefore given in the $|R\rangle$, $|L\rangle$-basis as

$$N = \begin{pmatrix} e^{-i\beta/2} & 0 \\ 0 & e^{i\beta/2} \end{pmatrix}$$

(a) What is the matrix N' of the same circular retarder in the $|R\rangle$, $|L\rangle$-basis? *Hint*: Make an 'obvious' change in the matrix U.
(b) A P_x-state is incident on an optical active material with $\beta = 180°$. In what state is the output beam?

9.5 Equations (9.71) and (9.72) are derived in the $|P_x\rangle$, $|P_y\rangle$-basis. Do the same calculations in the $|R\rangle$, $|L\rangle$-basis.

9.6 The indices of refraction for calcite and quartz for the sodium yellow line are

$$\text{calcite: } n_0 = 1.658, \quad n_e = 1.486$$
$$\text{quartz: } n_0 = 1.544 \quad n_e = 1.553$$

Calculate the thickness of a quarterwave plate made from these materials.

9.7 Consider a quarterwave plate designed for wavelength λ_0. Discuss its behaviour for wavelengths λ near λ_0, that is for small $\varepsilon = (\lambda_0/\lambda - 1)$. Imagine the use of this plate as the 'photoelastic model' in the dark-field plane polariscope in Figure 9.9. When rotating the quarter-wave plate the output intensity will vary. Compute the amplitude of this variation for $\varepsilon = 0$ and $\varepsilon = 0.1$.

9.8 Show that the Brewster angles for internal and external reflection at a given interface are complementary, i.e. $\theta_{B1} + \theta_{B2} = 90°$.

9.9 Prove that for a plane parallel slab if Brewster's condition is satisfied at the top interface it will be satisfied also for internal incidence onto the second interface.

9.10 Show that when $\theta_1 > \theta_c$ at a dielectric interface, r_p and r_n are complex and $r_n r_n^* = r_p r_p^* = 1$.

9.11 Find an expression for the relative phase shift between u_{rp} and u_{rn} when $n_1 > n_2$. At which angle of incidence is the phase shift maximum for $\theta_1 > \theta_c$?

9.12 The distance travelled by an evanescent wave at total internal reflection to a depth where the intensity is $1/e$ of its initial value is called the 'skin depth' δ.

(a) Find an expression for δ.
(b) How many wavelengths is the skin depth for $\theta_1 = 45°$ when $n_1 = 1.5$, $n_2 = 1$.

10 Digital image processing

10.1 THE FRAME GRABBER

A continuous analogue representation (the video signal) of an image cannot be conveniently interpreted by a computer and an alternative representation, the digital image, must be used. It is generated by an analogue-to digital (A/D) converter, often referred to as a 'digitizer', a 'frame-store' or a 'frame-grabber'. With a frame grabber the digital image can be stored into the frame memory giving the possibility of data processing and display. The block diagram of a frame grabber module is shown in Figure 10.1. These blocks can be divided into four main sections:

(1) the video source interface;

(2) multiplexer and input feedback LUT;

(3) frame memory;

(4) display interface.

The video source interface

The video source interface performs three main operations: (1) signal conditioning, (2) synchronization/timing and (3) digitalization. In the signal condition circuitry the signal is low-pass filtered with a cut-off frequency of 3–5 MHz to avoid aliasing, see Section 5.7.3. Some frame grabbers also have programmable offset and gain.

Before A/D conversion, the video signal must be stripped from its horizontal and vertical sync signals. The pixel clock in the frame grabber defines the sampling interval of the A/D converter and generates an internal Hsync signal. A phase-locked loop (PLL) tries to match this Hsync with the Hsync signal from the camera by varying the pixel clock frequency. This matching is iterative so it takes some time before the Hsyncs fit. And even then this process keeps on oscillating and produce a phenomenon called line jitter. Line jitter is therefore a mismatch and a wrong sampling of the analogue signal and has its largest effect in the first TV lines (upper part of the picture). The error can be as high as one pixel and therefore may ruin measurements which aims at subpixel accuracy.

Most frame grabbers have an 8-bit flash converter but 12- and 16-bit converters exists. The converter samples the analogue video signal at discrete time intervals and converts each sample to a digital value called a pixel element or pixel. The incoming signal is an analogue signal ranging from 0 to 714 mV at a frequency range from 7 to 20 MHz (with no prefiltering). The 8-bit converter produces samples with intensity levels between 0 and 255, i.e. 256 different grey levels.

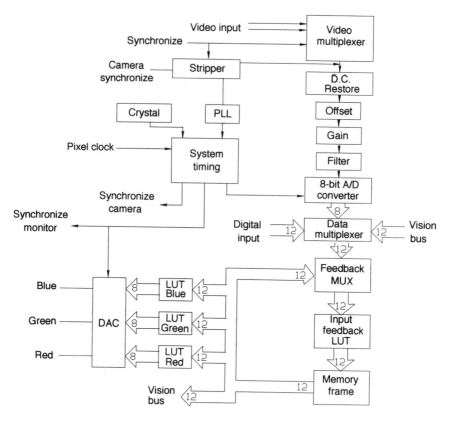

Fig. 10.1. *Block diagram of frame grabber*

Multiplexer and input feedback LUT

The frame memory of an 8-bit converter has a depth of 12 bits. The 4-bit spare allows the processor to use look-up table (LUT) operations. The LUT transforms image data before it is stored in the frame memory. The LUTs are mostly used for colouring (false colours) of images and can also be used to draw graphics on the screen without changing the underlying image. This can be done by protecting the lower 8 bits (the image) and draw only in the upper 4 bits. It is therefore possible to grab a new image without destroying the graphics. The LUT operations can be done in real time and its therefore possible to correct images radiometrically before storing them. LUTs can not be used geometrically because their memory is pixel organized, not space oriented. For geometrical transformations one therefore has to make special programs. The multiplexer (MUX) in combination with the feedback/input LUT allows some feedback operations like real-time image differencing, low pass filtering, etc.

Frame memory

The frame memory is organized as a two-dimensional array of pixels. Depending on the size of the memory it can store one or more frames of video information. When the memory is a 12-bit memory, 8 bits are used for the image and 4 bit planes for generating graphic overlay or LUT operations. In normal mode the memory acquires and displays an image using read/modify/write cycles. The memory is XY addressed to have an easy and fast access to single pixels.

Display interface

The frame memory transports the contents to the display interface every memory cycle. The display interface transforms the digital 12-bit signal from the frame memory into an analogue signal with colour information. This signal is passed to the RED, GREEN and BLUE ports and from there to the monitor.

10.2 DIGITAL IMAGE REPRESENTATION

By means of an electronic camera and a frame grabber, an image will be represented as a two-dimensional array of pixels, each pixel having a value $g(x, y)$ between 0 and 255 representing the grey tone of the image in the pixel position. Most current commercial frame grabbers have an array size of 512×512 pixels. Due to the way the image is scanned, the customary XY coordinate axis convention is as indicated in Figure 10.2.

10.3 CAMERA CALIBRATION

The calibration of the camera/lens combination is the process of determining the correct relationships between the object and image coordinates (Tsai 1987; Lenz and Tsai 1988). Since the elements of such a system are not ideal, this transformation includes parameters that must be calibrated experimentally. Because we are mainly concerned with relative measurements, we confine our discussion to three parameters that will affect our type of

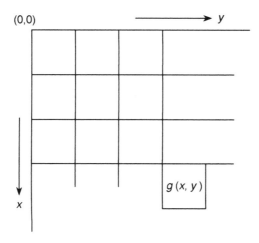

Fig. 10.2. *Digital image representation*

measurements. That is lens distortion, image centre coordinates and perspective transformations.

10.3.1 Lens distortion

For an ideal lens, the transformation from object (x_o, y_o) to image (x_i, y_i) coordinates is
simply

$$x_i = mx_o \tag{10.1a}$$

$$y_i = my_o \tag{10.1b}$$

where m is the transversal magnification. It is well known, however, that real lenses
possesses distortion to a smaller or larger extent (Faig 1975; Shih *et al.* 1993). The transfer
from object to image coordinates for a distorting lens is (see Figure 10.3)

$$r_i = mr_o + d_1 r_o^3 \tag{10.2a}$$

where

$$r_i = \sqrt{x_i^2 + y_i^2}, \quad r_o = \sqrt{x_o^2 + y_o^2} \tag{10.2b}$$

Higher odd order terms of r_o may be added, but normally they will be negligible. By
multiplying Equation (10.2a) by $\cos\phi$ and $\sin\phi$, we get (since $x = r\cos\phi, y = r\sin\phi$)

$$x_i = mx_o + d_1 x_o(x_o^2 + y_o^2) \tag{10.3a}$$

$$y_i = my_o + d_1 y_o(x_o^2 + y_o^2) \tag{10.3b}$$

This results in the well-known barrel (positive d_1) and pin-cushion (negative d_1) distortion.

In a digital image-processing system we want to transform the distorted coordinates
(x_d, y_d) to undistorted coordinates (x_u, y_u). This transformation becomes

$$x_u = x_d + dx_d(x_d^2 + \varepsilon^2 y_d^2) \tag{10.4a}$$

$$y_u = y_d + dy_d(x_d^2 + \varepsilon^2 y_d^2) \tag{10.4b}$$

where ε is the aspect ratio between the horizontal and vertical dimensions of the pixels.

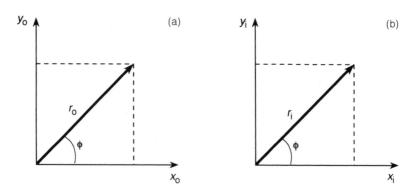

Fig. 10.3. *(a) Object and (b) image coordinates*

The origin of the xy-coordinates is at the optical axis. When transforming to the frame-store coordinate system XY (see Section 10.2) by the transformation

$$x = X - X_s \tag{10.5a}$$

$$y = Y - Y_s \tag{10.5b}$$

Equation (10.4) becomes

$$X_u = X_d + d(X_d - X_s)[(X_d - X_s)^2 + \varepsilon^2(Y_d - Y_s)^2] \tag{10.6a}$$

$$Y_u = Y_d + d(Y_d - Y_s)[(X_d - X_s)^2 + \varepsilon^2(Y_d - Y_s)^2] \tag{10.6b}$$

where (X_s, Y_s) are the centre coordinates. The distortion factor d has to be calibrated by e.g. recording a scene with known, fixed points or straight lines. The magnitude of d is of the order of 10^{-6}–10^{-8} pixels per mm^3.

It has been common practice in the computer vision area to choose the center of the image frame buffer as the image origin. For a 512×512 frame buffer that means $X_s = Y_s = 255$. With a CCIR video format, the center coordinates would rather be $(236,255)$ since only the first 576 of the 625 lines are true video signals, see Table 5.6. A mispositioning of the sensor chip in the camera could add further to these values. The problem is then to find the coordinates of the image center. Many methods have been proposed, one which uses the refection of a laser beam from the frontside of the lens (Tsai 1987). When correcting for camera lens distortion, correct image center coordinates are quite important.

10.3.2 Perspective transformations

Figure 10.4 shows a lens with the conjugate object- and image planes and with object and image distances a and b respectively. A point with coordinates (x_p, z_p) will be imaged (slightly out of focus) with image coordinate $-x_i$, the same as for the object point $(x_o, 0)$. From similar triangles we find that

$$x_p = \frac{-x_i(z_p + a)}{b} \tag{10.7a}$$

$$y_p = \frac{-y_i(z_p + a)}{b} \tag{10.7b}$$

Equation (10.7) is the perspective transformation and must be taken into account when e.g. comparing a real object with an object generated in the computer.

10.4 IMAGE PROCESSING

Broadly speaking, digital image processing can be divided into three distinct classes of operations: point operations, neighbourhood operations and geometric operations. A point operation is an operation in which the grey level of each pixel in the output image is a function of the grey level of the corresponding pixel in the input image, and only of that pixel. Typical point operations are photometric decalibration, contrast stretching and thresholding. A neighbourhood operation generates an output pixel on the basis of the

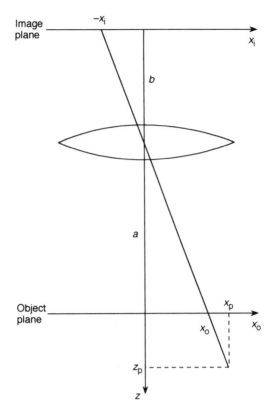

Fig. 10.4. *Perspective transformation*

grey level of the corresponding pixel in the input image and its neighbouring pixels. Geometric operations change the spatial relationships between points in an image, i.e. the relative distances between points a, b and c will typically be different after a geometric operation or 'warping'. Correcting lens distortion is an example of geometric operations. Digital image processing is a wide and growing topic with an extensive literature (Vernon 1991; Baxes 1984; Niblack 1988; Gonzales and Wintz 1987; Pratt 1991; Rosenfeld and Kak 1982). Here we'll treat only a small piece of this large subject and specifically consider operation, which can be very useful for enhancing interferograms, suppress image noise, etc.

10.4.1   Contrast stretching

In a digitized image we may take the number of pixels having the same grey level and make a plot of this number of pixels as a function of grey level. Such a plot is called a grey-level histogram. For an 8-bit (256 grey levels) image it may look such as that in Figure 10.5(a). In this example the complete range of grey levels is not used and the contrast of this image will be quite poor. We wish to enhance the contrast so that all levels of the grey scale are

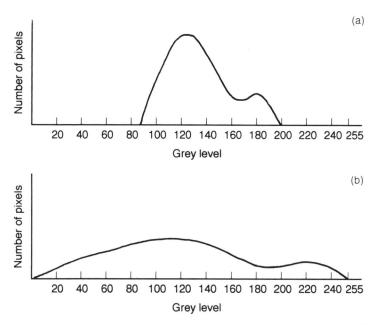

Fig. 10.5. *Grey-level histogram (a) before and (b) after contrast stretching*

utilized. If the highest and lowest grey value of the image are denoted g_H and g_L respectively, this is achieved by the following operation:

$$g_N = (g_o - g_L)\frac{255}{(g_H - g_L)} \tag{10.8}$$

where g_o is the original grey value and g_N is the new grey value. This is called contrast stretching and the resulting histogram when applied to the image of Figure 10.5(a) is given in Figure 10.5(b).

Grey-level histograms can be utilized in many ways. Histogram equalization is a technique which computes the histogram of an image and reassigns grey levels to pixels in an effort to generate a histogram where there are equally many pixels at every grey level, thereby producing an image with a flat or uniform grey-level histogram.

10.4.2 Neighbourhood operations. Convolution

Neighbourhood processing is formulated in the context of so-called mask operations (the terms template, window or filter are also often used to denote a mask). The idea behind mask operations is to let the value assigned to a pixel be a function of itself and its neighbours. The size of the neighbourhood may vary, but techniques using 3×3 or 5×5 neighbourhoods centred at the input pixel are most common. The neighbourhood operations are often referred to as filtering operations. This is particularly true if they involve the convolution of an image with a filter kernel or mask. Other neighbouring operations are concerned with modifying the image, not by filtering in the strict sense, but by applying some logical test based on e.g. the presence or absence of object pixels in the local neighbourhood surrounding the pixel in question. Object thinning or skeletonizing is a typical example of this type of operation, as are the related operations of erosion and dilatation, which respectively seek to contract and enlarge an object in an orderly manner.

Recall the two-dimensional convolution integral

$$g(x, y) = f(x, y) \otimes h(x, y) = \int_{-\infty}^{\infty} \int_{-\infty}^{\infty} f(\xi, \eta) h(x - \xi, y - \eta) \, d\xi \, d\eta \qquad (10.9)$$

When the variables x, y are not continuous but merely discrete values m, n as the pixel numbers in the x- and y-direction as of a digitized image, the double integral has to be replaced by a double sum:

$$g(i, j) = f \otimes h = \sum_{m} \sum_{n} f(m, n) h(i - m, j - n) \qquad (10.10)$$

As mentioned in Appendix B, the geometrical interpretation of the convolution $f \otimes h$ is the area of overlap between the functions f and h as a function of the position of h as h is translated from $-\infty$ to ∞. Therefore the summation is taken only over the area where f and h overlap. This multiplication and summation is illustrated graphically in Figure 10.6(a). Here the filter kernel h is a 3×3 pixel mask with the origin $h(0, 0)$ at the centre representing a mask of nine distinct weights, $h(-1, -1), \ldots, h(+1, +1)$, see Figure 10.6(b). Note that the convolution formula requires that the mask h be first rotated $180°$, but this can be omitted when the mask is symmetric.

10.4.3 Noise suppression

High spatial frequencies in an image manifests itself as large variations in grey tone values from one pixel to the next. This may be due to unwanted noise in the image. Such variations can be smoothed out by a convolution operation using an appropriate mask, the mask values constitute the weighting factors which will be applied to the corresponding image point when the convolution is being performed. For example, each of the mask values might be equally weighted, in which case the operation is simply the evaluation of the local mean of the image in the vicinity of the mask. Such a mask therefore acts as a low-pass filter. Figure 10.7(a) shows this local neighbourhood average mask and Figure 10.7(b) illustrates the application of the mask to part of an image. Referring to Figure 10.7(b) we find the value of the output pixel which replace the input pixel corresponding to the centre position of the mask to be

$$1/9[102 + 105 + 100 + 109 + 142 + 120 + 130 + 128 + 126] = 118$$

Thus the central point becomes 118 instead of 142 and the image will appear much smoother.

It may be more useful to apply this smoothing subject to some condition, e.g. the centre pixel is only assigned if the difference between the average value and the original pixel value is greater than a previously set threshold. This remove some of the noise without smoothing out too much of the detail in the original image.

(a)

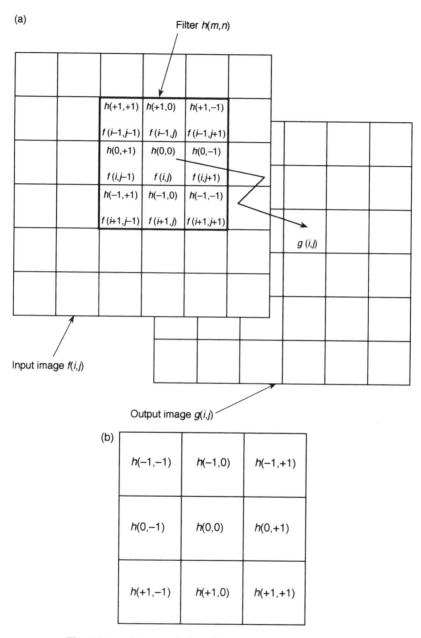

Fig. 10.6.   *(a) convolution; (b) 3 × 3 convolution filter h*

Of the other noise-suppression methods we mention median filtering. This is a noise-reduction technique whereby a pixel is assigned the value of the median in some local neighbourhood. This is, however, a computationally time consuming procedure so neighbourhoods in excess of 3 × 3 or 5 × 5 may be impractical. In general, the median filter is superior to the mean filter in that image blurring is minimized.

(a)

1 / 9	1 / 9	1 / 9
1 / 9	1 / 9	1 / 9
1 / 9	1 / 9	1 / 9

(b)

120 × 1 / 9	105 × 1 / 9	100 × 1 / 9				
109 × 1 / 9	142 × 1 / 9	120 × 1 / 9				
130 × 1 / 9	128 × 1 / 9	126 × 1 / 9				

Fig. 10.7. *(a) 3 × 3 average mask; (b) image smoothing using local average mask*

10.4.4 Edge detection

Detection of edges of objects in an image is a fundamental problem, not only for dimensional measurements, but in object recognition, military surveillance, etc. In the latter applications the main concern is if an edge exists or not, while in optical metrology one is also interested in the exact location of the edge. A lot of different edge detection techniques are described in the literature (Roberts 1965; Prewitt 1970; Fram and Deutsch 1975; Canny 1986). The different approaches can be divided into:

(1) gradient- and difference-based operators;

(2) template matching;

(3) edge fitting;

(4) statistical edge detection.

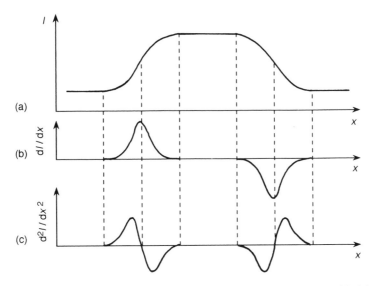

Fig. 10.8. *(a) intensity distribution I across two edges, (b) the derivative of I. (c) the second derivative of I*

Here we will not treat all of these methods, but confine ourselves to describe in some detail the 'classical' method based on gradient- and difference-based operators.

An image of an edge will result in an intensity distribution consisting of a step-like function, see Figure 10.8(a). The actual position of the edge might be questioned but it is commonly adopted that the position is where the gradient has its maximum absolute value, see Figure 10.8(b) (if the edge transition is from bright to dark, the derivative becomes negative). So how do we calculate the gradient in a digitized image? From the definition of the derivative we have

$$f'(x) = \lim_{\Delta x \to 0} \frac{f(x + \Delta x) - f(x)}{\Delta x} \tag{10.11}$$

In an image of distinct pixel values the closest we can come to the limit $\Delta x \to 0$ is unity. The best approximation of the (partial) derivative in the x-direction we can give is therefore simply

$$S_x = f(x + 1, y) - f(x, y) \tag{10.12a}$$

and in the y-direction

$$S_y = f(x, y + 1) - f(x, y) \tag{10.12b}$$

The gradient S is a vector

$$S = S_x e_x + S_y e_y \tag{10.13}$$

of magnitude

$$S = \sqrt{S_x^2 + S_y^2} \tag{10.14a}$$

and a direction given by

$$\tan\theta = \frac{S_y}{S_x} \qquad (10.14b)$$

The gradient magnitude and direction can be found from the two components along any two orthogonal directions.

The essential difference between all gradient-based edge detectors are the directions which the operators use, the manners in which they approximate the one-dimensional derivatives along these directions, and the manner in which they combine these approximations to form the gradient magnitude.

An operator due to Roberts estimates the derivatives diagonally over a 2×2 neighbourhood and the gradient is approximated by

$$S(x, y) = \sqrt{[f(x, y) - f(x + 1, y + 1)]^2 + [f(x, y + 1) - f(x + 1, y)]^2} \qquad (10.15)$$

One of the main problems with the Roberts operator is its susceptibility to noise because of the manner in which it estimates the directional derivatives. This has prompted alternative estimations of the gradient by combining the differencing process with local averaging. For example, the Sobel operator estimates the partial derivative in the x- and y-direction over a 3×3 region centred at $f(x, y)$ by:

$$S_x = [f(x + 1, y - 1) + 2f(x + 1, y) + f(x + 1, y + 1)]$$
$$- [f(x - 1, y - 1) + 2f(x - 1, y) + f(x - 1, y + 1)] \qquad (10.16a)$$

$$S_y = [f(x - 1, y + 1) + 2f(x, y + 1) + f(x + 1, y + 1)]$$
$$- [f(x - 1, y - 1) + 2f(x, y - 1) + f(x + 1, y - 1)] \qquad (10.16b)$$

The gradient may then be estimated as before by either calculating the RMS

$$S = \sqrt{S_x^2 + S_y^2} \qquad (10.17)$$

or by taking the absolute values

$$S = |S_x| + |S_y| \qquad (10.18)$$

From Figure 10.9 we see that Equation (10.16a) is obtained by applying the convolution mask illustrated in Figure 10.9(b) (left) and similarly Equation (10.16b) is obtained by applying the mask in Figure 10.9(b) (right). The same figure also gives the convolution masks for the Roberts and Prewitt edge operators.

After applying the convolution mask we get an intensity distribution as sketched in Figure 10.8(b). To determine the position of the edge with subpixel accuracy, one of the techniques described in Section 11.2.4 can be applied. A simple but reliable solution is to use the method of fitting a quadratic curve to the three points $f(i - 1)$, $f(i)$ and $f(i + 1)$ where pixel i is the pixel of the highest intensity $f(i)$. Then the location of the maximum of the quadratic curve is given by

$$i + \frac{f(i + 1) - f(i - 1)}{2[(f(i) - f(i - 1)) - (f(i + 1) - f(i))]} \qquad (10.19)$$

To avoid detecting false edges, the magnitude of $f(i)$ can be compared to a predetermined threshold value and edges excluded with a maximum derivative below this

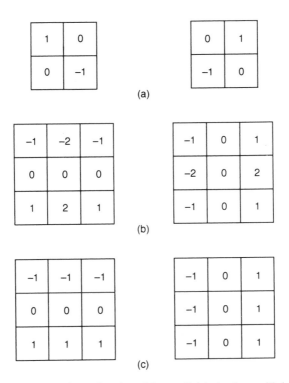

Fig. 10.9. *Convolution masks for estimation of the partial derivatives with (a) Roberts, (b) Sobel and (c) Prewitt edge detection operators*

threshold. Obviously the choice of threshold is important and in noisy images, threshold selection involves a trade-off between missing valid edges and including noise-induced false edges.

Finally we mention that edges can be detected using the second derivative by using the Laplacian:

$$\nabla^2 = \frac{\partial^2}{\partial x^2} + \frac{\partial^2}{\partial y^2} \tag{10.20}$$

i.e. the sum of second-order, unmixed partial derivatives. The standard approximation is given by

$$L(x, y) = f(x, y) - \tfrac{1}{4}[f(x, y+1) + f(x, y-1) + f(x+1, y) + f(x-1, y)] \tag{10.21}$$

From Figure 10.8(c) we see that the second derivative crosses zero at the position of the edge. Therefore the edge is detected by finding the points where the intensity goes from positive to negative. The Laplacian has, however, one significant disadvantage: it responds very strongly to noise. To reduce noise, the image can be smoothed. This was proposed by Marr and Hildreth (1980) by first smoothing the image by convolving it with a two-dimensional Gaussian function. This is however a rather computationally expensive operation.

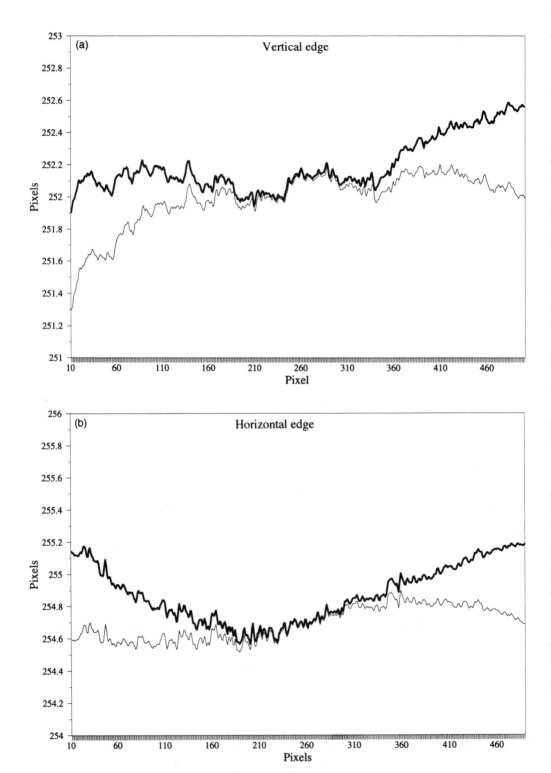

Fig. 10.10. *Detection of edge using the Sobel operator with sub-pixel accuracy. Thick curve before and thin curve after correcting for lens distortion: (a) vertical edge; (b) horizontal edge*

10.5   *EXPERIMENTAL EDGE DETECTION*

Figure 10.10 shows the result of an experiment for detecting a straight edge using an interlaced frame transfer CCD video camera. Here the edge is detected by use of the Sobel edge operator and sub-pixel accuracy is obtained by fitting three measured points around the intensity maximum to a quadratic curve. Figure 10.10(a, b) shows the result from detecting a vertical and a horizontal edge respectively. The two curves in each figure show the result before and after the data being corrected for camera lens distortion. In the left part of Figure 10.10(a) we see the effect of line jitter. It takes some time before the phase-locked loop (PLL) matches the internal Hsync signal. The left part of the corrected curve therefore will be overcompensated for lens distortion and gives an erroneous result. When the edge is horizontal as in Figure 10.10(b) a better result is obtained although the resolution in the vertical direction is poorer.

PROBLEMS

10.1 A common measure for the transmission of digital data is the baud rate, defined as the number of bits transmitted per second. Usually each byte (8 bits) is accompanied by a start bit and a stop bit. What then, is the transmit time for

(a) a 512×512 image with 256 grey levels at 300 baud?
(b) The same at 14 400 baud?
(c) Repeat (a) and (b) for a 1024×1024 image with 256 grey levels.

10.2 How does contrast stretching influence edge detection by a gradient-based operator?

10.2 Obtain the Fourier transforms of (a) $df(x)/dx$, (b) $[\partial f(x, y)/\partial x + \partial f(x, y)/\partial y]$, and (c) $[\partial^2 f(x, y)/\partial x^2 + \partial^2 f(x, y)/\partial y^2]$. Assume that x and y are continuous variables.

10.3 The basic approach used to compute the digital gradient involves taking differences of the form $f(x, y) - f(x + 1, y)$.

(a) Obtain the filter transfer function, $H(f_x, f_y)$, for performing the equivalent process in the frequency domain.
(b) Show that this is a high-pass filter.

10.4 The result obtained by a single pass through an image of some two-dimensional masks can also be achieved by two passes using one-dimensional masks.

(a) Show that the result obtained by a single pass of the mask given in Figure 10.7(a) also can be obtained by first passing the mask [1 1 1] through the image followed by a vertical counterpart.
(b) Show that the Sobel operator can be implemented by one pass of a differencing mask of the form $[-1\ 0\ 1]$ (or its vertical counterpart) followed by a smoothing mask of the form

$$\begin{pmatrix} 1 \\ 2 \\ 1 \end{pmatrix}$$

(or its horizontal counterpart).

(c) If the mask is represented by $h(x, y)$, what is the condition that it can be implemented with one horizontal one-dimensional mask followed by one vertical one-dimensional mask.

10.5 Write down the 3×3 mask representing the Laplacian operator given in Equation (10.21).

10.6 The two-dimensional Gaussian function is given by

$$G(x, y) = \frac{1}{2\pi\sigma^2} \exp[-(x^2 + y^2)/2\sigma^2]$$

An approach to edge detection (Marr and Hildreth) is by first smoothing the image by convolving it with a two-dimensional Gaussian function and subsequently isolating the zero-crossings of the Laplacian of this image:

$$\nabla^2\{I(x, y) \otimes G(x, y)\}$$

(a) Show that the two operations commute such that we can derive a single filter: the Laplacian of Gaussian.
(b) Show that this operation can be implemented by first an operation in the x-direction followed by an operation in the y-direction.

11 Fringe analysis

11.1 INTRODUCTION

In Chapters 3 and 6–9 we have given a description of classical interferometry, holographic interferometry, moiré, speckle and photoelasticity. The outcome of all these techniques is a set of fringes called interferograms. For many years, the analysis of these interferograms has been a matter of manually locating the positions and numbering of the fringes. With the development and decreasing cost of digital image processing equipment, a lot of effort has been made into what is termed digital fringe pattern measurement techniques. It is three main reasons for this effort: (1) to obtain better accuracy; (2) to increase the speed; (3) to automate the process.

In this chapter, some of the basic principles of digital fringe pattern analysis will be described. In Section 11.2 we describe techniques which intend to be a direct replacement of the human brain–eye combination by detecting the positions of the fringes. In Sections 11.3, 11.4 and 11.5, techniques for continuous determination of the phase of the fringe function are described. For more detailed discussions of digital fringe pattern measurement techniques, the book edited by Robinson and Reid (1993) is recommended. A comparison of the different techniques is performed by Kujawinska 1993b, Perry and McKelvie 1993.

11.2 INTENSITY-BASED ANALYSIS METHODS

11.2.1 Introduction

Before the development of the phase-measurement techniques described in Sections 11.3 and 11.4, intensity-based techniques were the only image-processing tools available for the automatic analysis of interferograms. They still are important methods in fringe analysis and are sometimes the only viable technique for interferograms which have been retrieved from photographic records or which have been obtained from interferometers in which it is impossible or impractical to introduce phase-measuring techniques. They may be appropriate simply because quantitative results are not even needed.

For intensity-based methods it is very important to minimize the influence of noise, including speckle noise. Therefore preprocessing of the interferograms is highly recommended. Techniques most commonly used and described in Section 10.4.3 are low-pass filtering and median filtering. Specially designed for fringe analysis are the so-called spin-filters (Yu *et al.* 1994). Another method (if possible) is to combine two interferograms of opposite phase. When an interferogram is shifted in phase by π radians, the resulting dark fringes occupy the previous location of bright fringes and vice versa. If the noise is stationary, the noise distribution is unaffected by this shift. Then if the two interferograms are subtracted, the noise will subtract to give zero, while the fringes will combine to give a higher contrast than in the original. When the noise has approximately the same spatial frequency as the fringes, this method might be the only workable one for noise reduction.

11.2.2 Prior knowledge

When interferometric techniques are used for repetitive calibration, non-destructive testing or inspection, it is sometimes possible to design a comparatively simple fringe analysis system. After identifying the characteristics of the fringe pattern which are peculiar to the application, the capabilities of the analysis system can then be confined to those required for the measurement in question.

A very simple analysis method can be applied to moiré technique using projected fringes where the shape of a manufactured component can be compared with that of a master component. For a manufactured component identical to the master, the resulting image becomes uniformly dark. As the manufactured component deviates from the master, fringes appear in the image and hence the total intensity of the image increases. Der Hovanesian and Hung (1982) used this technique to control component shape by limiting the total image intensity to a threshold value.

Many specialised fringe analysis procedures can be developed by utilizing prior knowledge of the fringe pattern. The Young fringe method (see Section 8.3.2) resulting in a set of parallel fringes is a typical example which has undergone a lot of investigations (Halliwell and Pickering (1993)). Another example is holographic interferometry applied

Fig. 11.1. *Procedure for the analysis of Figure 11.2 (From Robinson, D. W. (1983) Automatic fringe analysis with a computer image processing system, Applied Optics, 22, 2169–76. Reproduced by permission of The Optical Society of America and by Courtesy of D. W. Robinson)*

Fig. 11.2.   *Result of analysing a holographic interferogram by the method shown in Figure 11.1 (From Robinson, D. W. (1983) Automatic fringe analysis with a computer image processing system, Applied Optics, 22, 2169–76. Reproduced by permission of the Optical Society of America and by Courtesy of D. W. Robinson)*

to testing of honeycomb panels (Robinson (1983)). Debrazing of the honeycomb produces groups of nearly circular fringes of a particular size and fringe density. As shown in Figure 11.1 the procedure starts with counting the number of fringes along a number of horizontal lines through the image. When a comparatively large number of fringes appear on a given line, a flaw is presumed to exist on that line. Short vertical scans along the line are then used to search for the location of the flaw by looking for a comparatively large number of fringes in the vertical direction. Having identified the probable existence and location of a flaw, the system carries out a further check by counting fringes along each of four short vectors angularly spaced at 45°, centred on the probable flaw site. If the same number of fringes appear on each of the four vectors, then the existence of a flaw is confirmed. As shown in Figure 11.2 the flaw sites are marked with small crosses.

11.2.3   Fringe tracking and thinning

Boundary (contour) tracking and object thinning (skeletoning) are standard operations in digital image processing. In fringe analysis we talk about fringe tracking and thinning, which have a slightly different meaning. A number of fringe tracking and thinning procedures have been proposed.

Fringe tracking involves a search for the locus of the fringe maxima (or minima) by examining the pixel values in all directions from the starting point (often determined manually) and moving the pixel locus in the direction along which the sum of the intensity is maximized (or minimized) or alternatively the gradient is a minimum. (As opposed to

Fig. 11.3. *Holographic interferogram with a computer tracked fringe. (From Button, B. L., Cutts, J., Dobbins, B. N., Moxan, G. J. and Wykes, C. (1985) The identification of fringe positions in speckle patterns, Opt. and Lasers Technology, 17, 189–92. Reproduced by permission of Elsevier Science Ltd)*

boundary tracking where one searches for the maximum gradient value.) In this way only a limited set of the whole image array is examined (Button *et al.* 1985), see Figure 11.3.

Thinning and skeletonizing techniques use similar approaches to detect fringe peaks (or minima) but instead of following the peaks with a roving pixel locus, the whole image is subjected to a peak detection matrix. A procedure due to Yatagai (1982) uses two-dimensional peak detection, locally performed within a 5×5 pixel matrix as shown in Figure 11.4(a). With respect to the four directions shown in Figure 11.4(b), the peak conditions are defined as

$$P_{00} + P_{0-1} + P_{01} > P_{-21} + P_{-20}P_{-2-1}$$
$$P_{00} + P_{0-1} + P_{01} > P_{21} + P_{20} + P_{2-1}$$

$$(11.1)$$

for the x-direction, with similar expressions for the y-direction and the xy- and yx-directions. When the peak conditions are satisfied for any two or more directions, the object point is recognized to be a point on a fringe skeleton.

11.2.4 Fringe location by sub-pixel accuracy

By the methods described above, fringes are located with an accuracy of one pixel. We now describe two methods by which fringe positions can be determined with an accuracy of

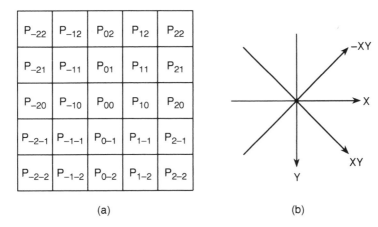

Fig. 11.4. *(a) 5 × 5 pixel matrix and (b) directions for fringe peak detection*

a fraction of a pixel. The first method consists off fitting a curve around the fringe maximum (or minimum)

Curve fitting (Chang et al. 1994)

Assume that the intensity distribution of the fringe pattern locally is given by

$$I(x) = a + b \cos \frac{2\pi}{p}(x - x_0) \tag{11.2}$$

with a maximum at $x = x_0$, i.e. $x_0 = np$ where n is an integer. Near the maximum we can approximate the intensity by a Taylor expansion of the intensity around x_0 up to the second order:

$$I(x) = a + b \left[1 - \frac{1}{2} \left(\frac{2\pi}{p} \right)^2 x^2 \right] \tag{11.3}$$

In Appendix C it is shown how such a quadratic curve can be fitted to N observation points by using the least squares solution. By using three observation points $I(i-1)$, $I(i)$ and $I(i+1)$ where i is the pixel number with the highest intensity $I(i)$, the position of the fringe maximum is given by, cf. Equation (C.19):

$$x_{max} = i + \frac{I(i-1) - I(i+1)}{2I(i-1) - 4I(i) + 2I(i+1)} \tag{11.4}$$

Thereby determining the position with sub-pixel accuracy.

The accuracy of this method is, among other things, dependent on the fringe period p. By using three measuring points, the length of p should at least be six pixels. If p is much longer than six pixels, it will be a good idea to fit the curve to four or more measurement points. It could also be advantageous to include four and higher-order terms in the expression for the intensity.

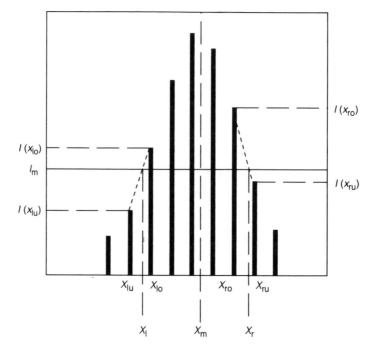

Fig. 11.5. *Illustration of the detection of the crossover points x_l and x_r with the mean intensity I_m*

Zero-crossing

Another method for sub-pixel location of fringe maxima (or minima) is to find the points where the intensity crosses the mean intensity value (Gåsvik *et al.* 1989). The principle is shown in Figure 11.5. On the left side of the maximum, the last pixel x_{lu} with intensity $I(x_{lu})$ below the mean intensity I_m and the first pixel x_{lo} with intensity $I(x_{lo})$ over I_m is found. Then a straight line

$$I = \frac{I(x_{lo}) - I(x_{lu})}{x_{lo} - x_{lu}} x + \frac{I(x_{lu})x_{lo} - I(x_{lo})x_{lu}}{x_{lo} - x_{lu}}$$

$$= [I(x_{lo}) - I(x_{lu})]x + [I(x_{lu}) - I(x_{lo})]x_{lu} + I(x_{lu}) \tag{11.5}$$

(The last equality follows since $x_{lo} - x_{lu} = 1$) connecting the two points are calculated. The intersection between this line and the mean intensity I_m is given by

$$x_l = \frac{I_m - I(x_{lu})}{I(x_{lo}) - I(x_{lu})} + x_{lu} \tag{11.6}$$

In the same way, the crossover point x_r on the right side of the maximum is found to be

$$x_r = \frac{I_m - I(x_{ru})}{I(x_{ru}) - I(x_{ro})} + x_{ru} \tag{11.7}$$

where x_{ro} is the last pixel with intensity $I(x_{ro})$ over I_m and x_{ru} is the first pixel with intensity

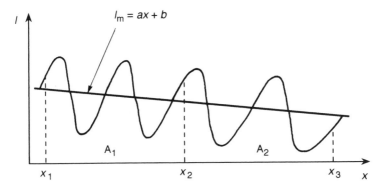

Fig. 11.6. *Intensity distribution and mean intensity I_m along a TV line*

$I(x_{ru})$ below I_m. The position x_m of the fringe maximum is then taken to be the midpoint

$$x_m = \frac{x_1 + x_r}{2} = \frac{1}{2}\left[\frac{I_m - I(x_{lu})}{I(x_{lo}) - I(x_{lu})} - \frac{I_m - I(x_{ru})}{I(x_{ro}) - I(x_{ru})}\right] + \frac{x_{lu} + x_{ru}}{2} \tag{11.8}$$

The mean intensity can be found in different ways. From Figure 11.6 we see that the area under the intensity curve and the area under the straight line

$$I_m = ax + b \tag{11.9}$$

representing the mean intensity should be approximately equal. If the interval from x_1 to x_3 is divided into two equal subintervals at x_2, we have

$$a = \frac{4(A_2 - A_1)}{N^2} \tag{11.10a}$$

$$b = \frac{3A_1 - A_2}{N} - \frac{4(A_2 - A_1)}{N^2}x_1 \tag{11.10b}$$

where $N = x_3 - x_1$, A_1 is the area of the first interval from x_1 to x_2 and A_2 is the area of the second interval from x_2 to x_3. These areas are found by simply taking the sum of the intensities for each pixel

$$A_1 = \sum_{i=x_1}^{x_2} I_i \tag{11.11a}$$

$$A_2 = \sum_{i=x_2}^{x_3} I_i \tag{11.11b}$$

To get the best possible representation of the mean intensity, each scan should be divided into as many intervals as possible, giving a sequence of straight lines representing I_m. Each subinterval must however not be shorter than one fringe period.

Another way of calculating the mean intensity is by the so-called bucket and bin method (Choudry 1981). It consists of finding the value of I_{min} and I_{max} (to the nearest integer pixel) for successive minima and maxima and then taking $I_m = (I_{max} + I_{min})/2$. For fringes with a high level of noise it could however be difficult to discriminate against spurious maxima and minima.

The accuracy of the zero-crossing method has been analysed with respect to the error in I_m and the fringe period (Gåsvik and Robbersmyr 1994). It shows that the absolute accuracy is highest for a fringe period of six pixels, while the relative accuracy is relatively constant for fringe periods of six pixels and higher. This analysis is done assuming noise-free fringes.

For more details about intensity-based analysis methods, see Yatagai 1993.

11.3 PHASE-MEASUREMENT INTERFEROMETRY

11.3.1 Introduction

By means of a digital image-processing system, we have the possibility of storing an image of the interferogram into the computer memory and do manipulations on the individual pixels. When looking at the general expression for an interferogram, Equation (3.7):

$$I = I_1 + I_2 + 2\sqrt{I_1 I_2}|\gamma|\cos \Delta\phi \tag{11.12}$$

it would be tempting to solve this expression with respect to $\Delta\phi$ and let the computer calculate:

$$\Delta\phi = \cos^{-1}\frac{I - (I_1 + I_2)}{2\sqrt{I_1 I_2}|\gamma|} \tag{11.13}$$

thereby calculating the phase at each pixel, and by knowledge of the geometrical and optical configuration of the interferometer, calculating the parameter sought in each pixel of the whole image. To do this, we see from Equation (11.13) that we must know the intensities I_1 and I_2 and the degree of mutual coherence $|\gamma|$ of the interfering waves, and, moreover, we must know these quantities for each pixel. This assumption is unrealistic in most cases, and even if we knew these quantities of the ideal interferogram, the intensity distribution from a complex imaging system will always be accompanied by uncontrollable noise. In this Section and in Sections 11.4 and 11.5 we will treat methods that are dependent only on the recorded intensity at each pixel and which are less sensitive to noise.

11.3.2 Principles of TPMI

A general expression for the recorded intensity in an interferogram can be written:

$$I(x, y) = a(x, y) + b(x, y)\cos \phi(x, y) \tag{11.14}$$

where both I, a, b and ϕ are functions of the spatial coordinates. Here $a(x, y)$ is the mean intensity, $V = a(x, y)/b(x, y)$ is equal to the visibility (see Section 3.3) and ϕ is the phase difference between the interfering waves.

To recover the phase we now describe a group of methods called phase-measurement interferometry (PMI). PMI is the most widely used technique today for the measurement of wavefront phase in interferometers and it has also been successfully applied in holographic interferometry and moiré (Rosvold 1990; Takeda and Mutoh 1983). PMI techniques can be divided into two main categories: those which take the phase data sequentially, and those which take the phase data simultaneously. Methods of the first type are known as temporal PMI or TPMI, and those of the second type are known as spatial PMI and will be treated in Section 11.4.

The starting point for all PMI techniques is the expression for the interferogram intensity:

$$I = a + b\cos(\phi + \alpha) \tag{11.15}$$

where we have introduced an additional phase term α. The essential feature of all PMI techniques is that α is a modulating phase which is introduced and controlled experimentally.

Techniques for determining the phase can be split into two basic categories: electronic and analytic. For analytical techniques, intensity data are recorded while the phase is temporally modulated, sent to a computer and then used to compute the relative intensity measurements. Electronic techniques are also known as heterodyne interferometry, see Section 3.5.4. An example of this technique is described in Section 3.5.3 about the dual-frequency Michelson interferometer. This method is used extensively in distance-measuring interferometers where the phase at a single point with a fast update is required. But the technique can also be used to determine the phase over an area, see Section 6.6.3. The detector then has to be scanned or there must be multiple detectors with all the necessary circuitry.

The analytic methods can be subdivided into two techniques, one that integrates the intensity while the phase is increased linearly, and a second where the phase is altered in steps between intensity measurements. The first method is referred to as integrating bucket phase-shifting, while the second is termed phase-stepping. The phase-step method has clearly become the most popular in recent years and below we give a brief description of this technique.

Equation (11.15) contains three unknowns, a, b and ϕ, requiring a minimum of three intensity measurements to determine the phase. The phase shift between adjacent measurements can be anything between 0 and π degrees. By arbitrary phase shifts α_1, α_2 and α_3 we get

$$\begin{aligned} I_1 &= a + b\cos(\phi + \alpha_1) \\ I_2 &= a + b\cos(\phi + \alpha_2) \\ I_3 &= a + b\cos(\phi + \alpha_3) \end{aligned} \tag{11.16}$$

from which we find

$$\phi = \tan^{-1}\frac{(I_2 - I_3)\cos\alpha_1 - (I_1 - I_3)\cos\alpha_2 + (I_1 - I_2)\cos\alpha_3}{(I_2 - I_3)\sin\alpha_1 - (I_1 - I_3)\sin\alpha_2 + (I_1 - I_2)\sin\alpha_3} \tag{11.17}$$

With $\alpha_1 = \pi/4$, $\alpha_2 = 3\pi/4$, $\alpha_3 = 5\pi/4$, i.e. a phase shift of $\pi/2$ per exposure, we reach a particularly simple expression

$$\phi = \tan^{-1}\frac{I_2 - I_3}{I_2 - I_1} \tag{11.18}$$

Three intensity measurements to solve for ϕ gives an exactly determined system. As mentioned in Appendix C, this often gives numerically unstable solutions and it is often wise to overdetermine the system by providing more measurement points.

In general, for the ith stepped phase, the resulting intensity can be written as (Greivenkamp 1984; Morgan 1982)

$$I_i = a + b\cos(\phi + \alpha_i) = a_0 + a_1\cos\alpha_i + a_2\sin\alpha_i \qquad (11.19a)$$

where

$$\begin{aligned} a_0 &= a = I_0 \\ a_1 &= b\cos\phi \\ a_2 &= -b\sin\phi \end{aligned} \qquad (11.19b)$$

By making N phase steps ($i = 1, 2, \ldots, N$), Equation (11.19a) can be written in matrix form as

$$\begin{pmatrix} I_1 \\ I_2 \\ \vdots \\ I_N \end{pmatrix} = \begin{pmatrix} 1 & \cos\alpha_1 & \sin\alpha_1 \\ 1 & \cos\alpha_2 & \sin\alpha_2 \\ \vdots & \vdots & \vdots \\ 1 & \cos\alpha_N & \sin\alpha_N \end{pmatrix} \begin{pmatrix} a_0 \\ a_1 \\ a_2 \end{pmatrix} \qquad (11.20)$$

The coefficients a_0, a_1 and a_2 can be found using the least squares solution to Equation (11.20), see Appendix C

$$\begin{pmatrix} a_0 \\ a_1 \\ a_2 \end{pmatrix} = A^{-1}B \qquad (11.21)$$

where

$$A = \begin{pmatrix} N & \Sigma\cos\alpha_i & \Sigma\sin\alpha_i \\ \Sigma\cos\alpha_i & \Sigma\cos^2\alpha_i & \Sigma\cos\alpha_i\sin\alpha_i \\ \Sigma\sin\alpha_i & \Sigma\cos\alpha_i\sin\alpha_i & \Sigma\sin^2\alpha_i \end{pmatrix} \qquad (11.22)$$

and

$$B = \begin{pmatrix} \Sigma I_i \\ \Sigma I_i\cos\alpha_i \\ \Sigma I_i\sin\alpha_i \end{pmatrix} \qquad (11.23)$$

From Equation (11.19b) we find that

$$\phi = \tan^{-1}\left(\frac{-a_2}{a_1}\right) \qquad (11.24)$$

$$V = \frac{b}{a} = \frac{\sqrt{a_1^2 + a_2^2}}{a_0} \qquad (11.25)$$

We include the expression for the visibility because pixels with too low visibility can give invalid phase data as will be discussed in Section 10.5.

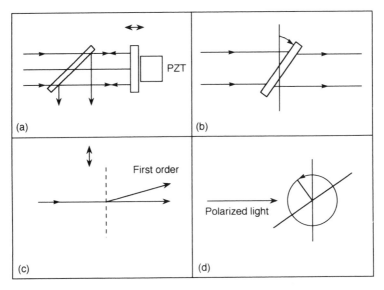

Fig. 11.7. *Means of modulating or shifting the phase of the light in an interferometer (a) moving mirror, (b) tilted glass plate; (c) moving diffraction grating; (d) rotating waveplate*

11.3.3 Means of phase modulation

A phase shift or modulation in an interferometer can be induced by moving a mirror, tilting a glass plate, moving a grating, rotating a half-wave plate or analyzer (see Figure 11.7) using an acousto-optic or electro-optic modulator, or using a Zeeman laser. Phase shifters such as moving mirrors, gratings, tilted glass plates, or polarization components (Jin *et al.* 1994) can produce continuous as well as discrete phase shifts between the object and reference beams. Phase shifters may either be placed in one arm of the interferometer or positioned so that they shift the phase of one of two orthogonally polarized beams.

11.3.4 Different techniques

From Equations (11.20)–(11.25) we can derive the expressions for different techniques, e.g.

Three-frame technique

(1) $\alpha_i = \pi/4, 3\pi/4, 5\pi/4$ (Wyant *et al.* 1984)

$$\phi = \tan^{-1}\left(\frac{I_3 - I_2}{I_1 - I_2}\right) \tag{11.26a}$$

$$V = \frac{\sqrt{(I_3 - I_2)^2 + (I_1 - I_2)^2}}{\sqrt{2}I_0} \tag{11.26b}$$

(2) $\alpha_i = -2\pi/3, 0, 2\pi/3$

$$\phi = \tan^{-1}\left[\sqrt{3}\frac{I_1 - I_3}{2I_1 - I_2 - I_3}\right] \tag{11.27a}$$

$$V = \frac{\sqrt{3(I_1 - I_3)^2 + (2I_2 - I_1 - I_3)^2}}{3I_0} \tag{11.27b}$$

Four-frame technique (Wyant 1982; Schwider et al. 1993)

$\alpha_i = 0, \pi/2, \pi, 3\pi/2$

$$\phi = \tan^{-1}\left(\frac{I_4 - I_2}{I_1 - I_3}\right) \tag{11.28a}$$

$$V = \frac{\sqrt{(I_4 - I_2)^2 + (I_1 - I_3)^2}}{2I_0} \tag{11.28b}$$

Five-frame technique (Hariharan et al. 1987; Schwider et al. 1983)

$\alpha_i = -\pi, -\pi/2, 0, \pi/2, \pi$

$$\phi = \tan^{-1}\left(\frac{7(I_2 - I_4)}{-4I_1 + I_2 + 6I_3 + I_4 - 4I_6}\right) \tag{11.29a}$$

$$V = \frac{\sqrt{(-4I_2 + I_2 + 6I_3 + I_4 - 4I_5)^2 + 49(I_2 - I_4)^2}}{14I_0} \tag{11.29b}$$

Hariharan uses slightly different formulas, not obtained by the least squares method.

Carré technique

In a technique due to Carré (1966) the amount of phase shift need not be known, but must be constant in the four steps, e.g. $\alpha_i = -3\alpha/2, -\alpha/2, \alpha/2, 3\alpha/2$. This gives

$$\alpha = 2\tan^{-1}\left(\sqrt{\frac{3(I_2 - I_3) - (I_1 - I_4)}{(I_2 - I_3) + (I_1 - I_4)}}\right) \tag{11.30a}$$

$$\phi = \tan^{-1}\left[\tan\left(\frac{\alpha}{2}\right)\frac{(I_1 - I_4) + (I_2 - I_3)}{(I_2 + I_3) - (I_1 - I_4)}\right] \tag{11.30b}$$

To calculate the phase modulo π, the above two equations are combined to yield

$$\phi = \tan^{-1}\frac{\sqrt{[(I_1 - I_4) + (I_2 - I_3)][3(I_2 - I_3) - (I_1 - I_4)]}}{(I_2 + I_3) - (I_1 + I_4)} \tag{11.31}$$

An obvious advantage of the Carré technique is that the phase does not need to be calibrated. It also has the advantage of working when a linear phase shift is introduced in a converging or diverging beam where the amount of phase shift varies across the beam. Unknown phase shift techniques with different number of steps are introduced by Vikram et al. 1993; Lassahn et al. 1994.

Ideally, two frames with a relative phase shift of 2π should be identical. This is the case for the five-frame technique where the phase difference between the first and the last frame is 2π. This 'extra' frame therefore can be used to correct for phase shifter miscalibration and detector non-linearities (Larkin and Oreb 1992). Techniques with a 2π phase difference between the first and the last frame are called $(N+1)$-frame techniques (Surrel 1993).

A technique due to Vikhagen (1990) requires the collection of many frames of intensity data with random phase shifts. From a number of recordings, the maximum and minimum intensity value at each detector point are determined. When the number of frames becomes large, these values will approach I_{max} and I_{min} which in turn are used to determine the mean intensity I_0 and the visibility V (see Equation 3.29). Once a large number of frames (about 20) of random phases has been recorded, the phase at each pixel of intensity I_i can be calculated according to

$$\phi = \cos^{-1}\left(\frac{I_i - I_0}{VI_0}\right) \tag{11.32}$$

11.3.5 Errors in TPMI measurements

The most common source of error in TPMI is vibration and air turbulence (Magee and Welsh 1994). In order to get good measurements, the optical system must be isolated from vibrations and shielded from air turbulence. Other factors that influence the measurement accuracy are phase-shifter errors, non-linearities due to the detector, quantization of the detector signal and TV-line jitter, see Section 10.1.

It has been shown that the errors in phase due to a calibration error or non-linearity in the phase shifter will decrease as the number of measurements increases (Schwider *et al.* 1983; Larkin and Oreb 1992; Joenathan 1994). For a consistent phase-shift error, such as miscalibration, a periodic error is seen in the calculated phase which has a spatial frequency of twice the fringe frequency. Figure 11.8 shows the peak-to-valley (P–V) phase

Fig. 11.8. *P–V phase error versus percent linear phase-shifter error. (From Creath 1993 in Robinson, D. W. and Reid, G. T. (eds) (1993) Interferogram Analysis. Digital Fringe Pattern Measurement Techniques, Institute of Physics Publishing. Reproduced by permission of Institute of Physics Publishing and by courtesy of K. Creath)*

error versus amount of linear phase-shifter error for a number of different algorithms (Creath 1993). As can be seen, the Carré and five-frame algorithms are the least sensitive to miscalibration.

Non-linear phase-shift errors are not as easy to detect or remove (Creath 1986), but also here the five-frame algorithm is the least sensitive to miscalibration.

With regard to detector non-linearities, the algorithms with four frames and more are quite insensitive to such errors. The three-frame algorithm, however, can cause large phase errors because of detector non-linearities.

For an intensity signal digitized to eight bits or more, the quantization error has an insignificant influence on the phase error.

For more details about TPMI techniques, see Creath (1993).

11.4 SPATIAL PHASE MEASUREMENT METHODS

11.4.1 Multichannel Interferometer

The temporal phase-stepping methods described in Section 11.3 has its spatial counterpart.

Consider Figure 11.9, which is a polarization (Michelson) interferometer. In the object and reference paths are placed two quarterwave plates (QWP) with their axes inclined at 45° to each other. The polarization matrix for a QWP with its fast axis inclined an angle ψ to the x-axis is (cf. Equation 9.61)

$$M_\psi = \begin{pmatrix} 1 - i\cos 2\psi & -i\sin 2\psi \\ -i\sin 2\psi & 1 + i\cos 2\psi \end{pmatrix} \tag{11.33}$$

The resultant matrix after two passes (first through the QWP and back after being

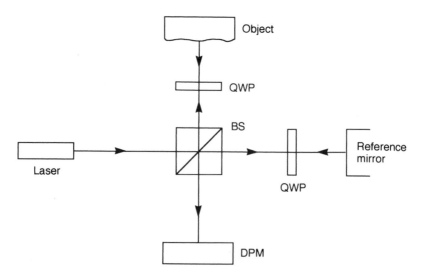

Fig. 11.9 *Polarization interferometer for direct phase measurement. BS = beamsplitter, QWP = quarter-wave plate, DPM = direct phase measurement module*

reflected from the mirror or object) therefore becomes

$$W_\psi = M_\psi M_\psi = \begin{pmatrix} -\cos 2\psi & -\sin 2\psi \\ -\sin 2\psi & \cos 2\psi \end{pmatrix} \tag{11.34}$$

which is the same as for a halfwave plate. The resultant matrix for a QWP with its axis inclined at 45° to that of Equation (11.34) becomes

$$W_{\psi+45} = \begin{pmatrix} \sin 2\psi & -\cos 2\psi \\ -\cos 2\psi & -\sin 2\psi \end{pmatrix} \tag{11.35}$$

Now assume that the state of polarization of the laser beam incident on the interferometer is represented by

$$|U\rangle = \begin{pmatrix} u_x \\ u_y \end{pmatrix} \tag{11.36}$$

If M_ψ is placed in the reference path and $M_{\psi+45}$ is placed in the object path, the state of polarization of the reference and object wave becomes

$$|U_r\rangle = W_\psi|U\rangle = \begin{pmatrix} u_{rx} \\ u_{ry} \end{pmatrix} = \begin{pmatrix} -u_x\cos 2\psi - u_y\sin 2\psi \\ -u_x\sin 2\psi + u_y\cos 2\psi \end{pmatrix} \tag{11.37a}$$

$$|U_o\rangle = W_{\psi+45}|U\rangle = \begin{pmatrix} u_{ox} \\ u_{oy} \end{pmatrix} = \begin{pmatrix} u_x\sin 2\psi - u_y\cos 2\psi \\ -u_x\cos 2\psi - u_y\sin 2\psi \end{pmatrix} \tag{11.37b}$$

The condition that these two states are orthogonal is, cf. Equations (9.46) and (9.48)

$$\langle U_r|U_o\rangle = \langle U_o|U_r\rangle = 0 \tag{11.38}$$

which becomes

$$u_x^* u_y - u_x u_y^* = 0 \tag{11.39}$$

This condition is fulfilled only when the incident laser beam is linearly polarized (u_x and u_y are real). Therefore, when placing two QWPs in the object- and reference paths with their axes inclined at 45° to each other and the incident beam is linearly polarized, the state of polarization of the resultant object- and reference waves becomes orthogonal (and linear). In this analysis we have neglected changes in the state of polarization due to reflections from the beamsplitter and possible depolarization by scattering from the object.

The light from the interferometer is directed to a direct phase measurement module (DPM). A three channel DPM (Bareket 1985) is illustrated in Figure 11.10. A camera imaging the interferogram is placed in each channel. In front of the camera is a linear polarizer with its transmission axis at 45° to the polarization direction of the object and reference waves, thereby mixing the two components. QWPs oriented parallel to the polarization direction shift the relative phase by $\pi/2$ an π in channel 1 and 2 respectively. The detected intensities in the three channels therefore become

$$\begin{aligned} I_1 &= a + b\cos(\phi + \pi/2) = a - b\sin\phi \\ I_2 &= a + b\cos(\phi + \pi) = a - b\cos\phi \\ I_3 &= a + b\cos\phi \end{aligned} \tag{11.40}$$

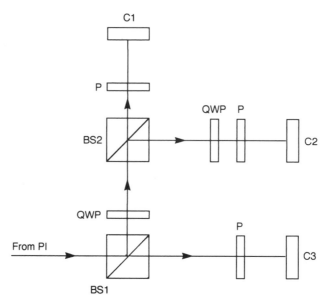

Fig. 11.10. *Direct phase measurement module DPM with three interference channels in parallel. PI = polarization interferometer, C1–C3 = cameras*

from which we solve

$$\phi = \tan^{-1}\left(\frac{2I_1 - I_2 - I_3}{I_2 - I_3}\right) \tag{11.41}$$

The DPM in Figure 11.10 can be extended further to four channels (Smythe and Moore 1984) or more and retarders with phase shifts other than $\pi/2$ can be inserted instead of the QWPs to get different types of spatial phase-stepping methods in analogy with the different types of temporal phase-stepping methods described in Section 11.3.4. But why introduce this technique with the rather complicated set-up described in Figure 11.9 and 11.10? The answer is that we are in this way able to record many phase-shifted interferograms simultaneously, thereby allowing analysis of dynamic events and pulsed wavefront sensing. It enables analysis in one frame time and therefore the measurements may be performed in adverse conditions.

Spatial phase-stepping can also be obtained by means of gratings (Kwon and Shough 1985). The transmission function of a square grating can be written as

$$t(x, y) = \sum_n a_n e^{i2\pi n f x} \tag{11.42}$$

where a_n are the coefficients of the Fourier series of the grating, f is the frequency, $f = 1/d$, and d is the grating period. A lateral translation of the grating by x_o results in an additional phase term described by

$$t(x - x_o, y) = \sum_n a_n e^{i2\pi n f(x - x_o)} \tag{11.43}$$

If a wavefront $U_o e^{i\phi}$ is incident on the grating, the field behind the grating is

$$u = U_o e^{i\phi} \sum_n a_n e^{i2\pi n f(x - x_o)} \tag{11.44}$$

In this way a series of phase-shifted wavefronts propagate behind the grating. The magnitude of the phase shift is given by

$$\delta_n = 2\pi n f x_o \tag{11.45}$$

Therefore, a three-channel interferometer can be obtained from the $n = +1, 0, -1$ orders of the diffracted beams. For $90°$ phase shift, the amount of lateral translation of the grating equals $d/4$. Grating phase-stepping has been successfully applied in two basic configurations. That is in (1) three-channel interferometers: point-diffraction (PDI) (Kwon and Shough 1985), radial-shear (RSI) (Kwon et al., 1987), holographic (Kujawinska and Robinson 1988) and speckle (Kujawinska et al. 1989) interferometers, and (2) four-channel grating lateral shear interferometers (Kwon and Shough 1985).

Apart from polarization and grating methods, also a colour technique in projection moiré has been demonstrated as a means for spatial phase-stepping (Harding et al. 1988).

11.4.2 Errors in multichannel interferometers

Multichannel interferometers are susceptible to errors due to conditions that are not common to all channels, such as are associated with using separate detectors. Errors may result from variations of the optical transmission among channels, detector responsitivity variations and amplifier gain and bias errors. These can, however, be adjusted either optically, electronically or in software. Another source of errors is connected with pixel mismatch between the fringe patterns due to incorrect relative spatial adjustment of the cameras. The overall accuracy of the spatial phase-stepped techniques is usually expected to be lower than for the temporal methods.

11.4.3 Spatial-carrier phase-measurement method

Spatial-carrier phase measurement methods are based on the idea of superposing a carrier fringe pattern onto the interferogram fringes. This can be done in e.g. holographic interferometry by tilting the reference wave in the second exposure. The fringe pattern is then given by

$$g(x, y) = a(x, y) + b(x, y) \cos[\phi(x, y) + 2\pi f_0 x] \tag{11.46}$$

where f_0 is the carrier frequency in the x-direction. The two basic approaches to the spatial carrier technique include the Fourier transform method (FTM), in which the processing is performed in the frequency domain, and methods which are equivalent but the processing is performed in spatial coordinates.

The Fourier transform method was originally conceived and demonstrated by Takeda et al. (1982) who employed the fast fourier transform (FFT) method. Following Takeda's method we write Equation (11.46) in the form

$$g(x, y) = a(x, y) + c(x, y) e^{i2\pi f_0 x} + c^*(x, y) e^{-2\pi f_0 x} \tag{11.47a}$$

where

$$c(x, y) = \tfrac{1}{2} b(x, y) e^{i\phi(x, y)} \tag{11.47b}$$

The fringe pattern is Fourier transformed with respect to x, which gives

$$G(f_x, y) = A(f_x, y) + C(f_x - f_0, y) + C^*(f_x + f_0, y) \tag{11.48}$$

where the upper-case letters denote Fourier spectra and f_x is the spatial frequency in the x-direction. To have the method work, the spatial variations of $\phi(x, y)$ must be slow compared to f_0 and the Fourier spectra will be separated as shown schematically in Figure 11.11(a). By use of a filter function $H(f_x - f_0, y)$ in the frequency plane the function $C(f_x - f_0, y)$ can be isolated and translated by f_0 towards the origin to remove the carrier and obtain $C(f_x, y)$ as shown in Figure 11.11(b). Next the inverse Fourier transform of this function is computed and as a result the complex function $c(x, y)$ from Equation (11.47b) is obtained. The phase may then be determined by two equivalent operations. In the first one a complex logarithm of $c(x, y)$ is calculated

$$\log[c(x, y)] = \log[1/2b(x, y)] + i\phi(x, y) \tag{11.49}$$

The phase in the imaginary part is completely separated from the amplitude variation $b(x, y)$ in the real part. In the second method (more commonly used) the phase is obtained by

$$\phi(x, y) = \tan^{-1} \frac{\text{Im}[c(x, y)]}{\text{Re}[c(x, y)]} \tag{11.50}$$

where Re and Im represent the real and imaginary parts of $c(x, y)$.

Figure 11.12 (Takeda et al. 1982) shows the various stages in the Fourier transform process. Before computing the FFT the data are multiplied by a Hamming window (Figure 11.12b) to eliminate the influence of discontinuities in the data at both ends.

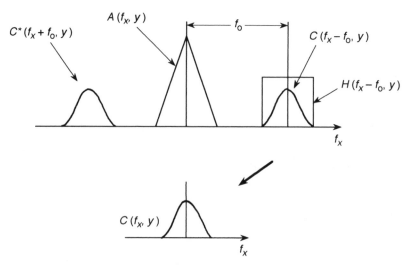

Fig. 11.11. *Separated Fourier spectra of a tilted fringe pattern*

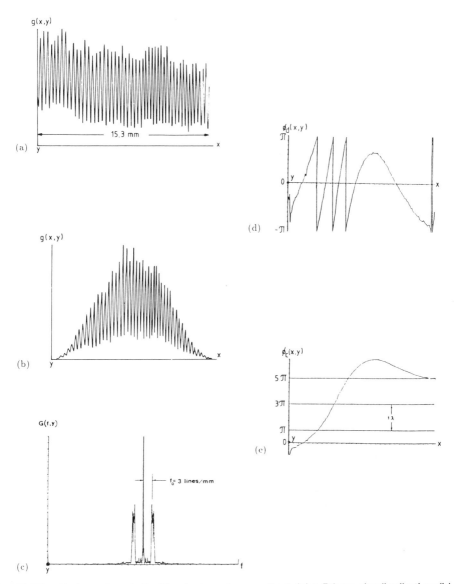

Fig. 11.12.   *Various stage in the Fourier transform method: (a) 1-D intensity distribution; (b) the intensity weighted by Hamming window; (c) the modulus of its Fourier transform; (d) the wrapped phase function; (e) the reconstructed phase function after the unwrapping procedure. (From Takeda, M., Ina, H. and Kobayashi, S. (1982) Fourier-transform method of fringe pattern analysis for computer-based topography and interferometry, J. Opt. Soc. Am., 72(1), 156–160. Reproduced by permission of the Optical Society of America and by courtesy of M. Takeda)*

The Fourier transform method has also been extended using a full two-dimensional spatial filtering (Nugent 1985; Bone *et al.* 1986).

The Fourier transform method has become quite popular and has also been applied to moiré grating techniques (Morimoto and Fujisawa 1994; Morimoto *et al.* 1994; Simova and Stoev 1993; Yoshizawa and Tomisawa 1993) and photoelasticity (Quan *et al.* 1993).

11.4.4 Errors in the Fourier transform method

Many of the error sources in the FTM are common to Fourier transform methods in general and have been studied extensively by many authors. The most serious source of error is energy leakage. A simplified explanation of this phenomenon goes as follows.

The FFT has to be computed over a sample window of finite width D, i.e. the function $g(x)$ of Equation (11.46) is in effect multiplied by a rect-function:

$$g_s(x) = g(x)\,\text{rect}\left(\frac{x}{D}\right) \tag{11.51}$$

and the Fourier transform gives

$$\mathscr{F}\{g_s\} = G(f_x, y) \otimes D\,\text{sinc}(Df_x) \tag{11.52}$$

From this expression we see that the spectrum $G(f_x, y)$ is convolved with a sinc-function. Due to the side-lobes of the sinc-function, parts of the spectrum of $G(f_x, y)$ will 'leak out' into neighbouring frequency components because of this convolution operation, resulting in a false spectrum. When multiplying the data with a Hamming window as proposed by Takeda *et al.*, the steep discontinuities at the boundaries of the fringe data are avoided, see Figure 11.12(b). The Hamming window is given by

$$w(x) = 1 - \cos\frac{2\pi}{D}x \tag{11.53}$$

and its Fourier transform

$$W(f_x) = \mathscr{F}\{w(x)\} = \delta(f_x) - \tfrac{1}{2}\delta\left(f_x - \frac{1}{D}\right) - \tfrac{1}{2}\delta\left(f_x + \frac{1}{D}\right) \tag{11.54}$$

When convolving this function with the spectrum $G(f_x, y)$ we get no leakage. The Hamming window is most commonly used, however Hann, bell and $\cos^4$ windows also give good results (Malcolm *et al.* 1989; Frankowski *et al.* 1989). However, by using such windows, another source of error comes up. Because the fringe amplitude approaches zero at the ends of the window (see Figure 11.12(b)), the determination of the phase becomes quite sensitive to noise in these regions.

11.4.5 Space domain

A process equivalent to the Fourier transform method can be performed in the space domain. From Equation (11.48) the spectrum passed by the filter function $H(f_x - f_0, y)$ is written as

$$C(f_x - f_0, y) = H(f_x - f_0, y)G(f_x, y) \tag{11.55}$$

By taking the inverse Fourier transform of Equation (11.55) we get

$$c(x, y)e^{i2\pi f_0 x} = g(x, y) \otimes h(x, y)e^{i2\pi f_0 x} \tag{11.56}$$

Many different approaches to the solution of Equation (11.56) have been proposed

(Womack 1984 (Singh and Sirkis 1994)). A solution by means of a dedicated hardware system has been proposed and (fully) demonstrated in the pioneering work by Ichioka and Inuiya (1972). The intensity given by Equation (11.46) is transformed into the following time varying electrical signal by the video tube:

$$g(t) = a(t) + b(t) \cos[\omega t - \phi(t)] \tag{11.57}$$

The procedure is then exactly the same as for synchroneous demodulation of a modulated carrier signal, see Section 3.5.4, where the signal is divided into two signals and multiplied by $\cos \omega t$ and $\sin \omega t$:

$$g(t) \cos \omega t = \frac{b}{2} \cos \phi + a \cos \omega t + \frac{b}{2} \cos(2\omega t - \phi) \tag{11.58a}$$

$$g(t) \sin \omega t = \frac{b}{2} \sin \phi + a \sin \omega t + \frac{b}{2} \sin(2\omega t + \phi) \tag{11.58b}$$

By low-pass filtering of these two signals, one is left with

$$C(t) = \frac{b}{2} \cos \phi \tag{11.59a}$$

$$S(t) = \frac{b}{2} \sin \phi \tag{11.59b}$$

from which the phase can be computed electronically:

$$\phi(t) = \tan^{-1}\left(\frac{S(t)}{C(t)}\right) \tag{11.60}$$

For more details about spatial PMI techniques, see Kujawinska (1993).

11.5 PHASE UNWRAPPING

11.5.1 Introduction

As mentioned in Section 11.3.2 a general expression for the intensity in an interferogram can be written

$$I = a + b \cos \phi \tag{11.61}$$

In Sections 11.3 and 11.4 we have described techniques for solving Equation (11.61) by collecting extra information from the interferometer to solve the sign ambiguity and reduce the influence of stationary noise. All these methods result in an equation of the form

$$\phi = \tan^{-1}\left(\frac{C}{D}\right) \tag{11.62}$$

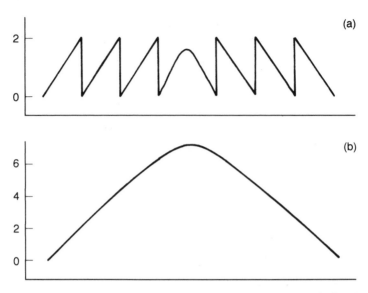

Fig. 11.13. *(a) Characteristic 'saw-tooth' wrapped phase function. (b) Continuous phase function obtained by 'unwrapping' the data in (a)*

where C and D are functions of the recorded intensity from a set of interferograms at the image point where the phase is being measured. Because of the multivalued arctan function, the solution for ϕ is a sawtooth function (see Figure 11.13(a)) and discontinuities occur every time ϕ changes by 2π. If ϕ is increasing, the slope of the function is positive and vice versa for decreasing phase. The term 'phase unwrapping' arises because the final step in the fringe pattern measurement process is to unwrap or integrate the phase along a line (or path) counting the 2π discontinuities and adding 2π each time the phase angle jumps from 2π to zero or subtracting 2π if the change is from zero to 2π. Figure 11.13(b) shows the data in Figure 11.13(a) after unwrapping.

The key to reliable phase-unwrapping algorithms is the ability to accurately detect the 2π phase jumps. In the case of noise-free wrapped phase data and where this data is adequately sampled (i.e. the phase gradients is significantly less than 2π), then a simple approach to phase unwrapping will be adequate and all that is required is a sequential scan through the data (line by line) to integrate the phase by adding or subtracting 2π at the phase jumps.

In many measurement problems, however, noise in the sampled data is a major contributing factor in the false identification of phase jumps. Figure 11.14 shows the effect of the addition of noise to unwrapped data. It is clear that as the amplitude of the noise approaches 2π, the real phase jumps become obscured. In the case of one-dimensional data, the only solution to this problem (other than averaging more data sets to integrate out the noise over time) is to smooth the raw sinusoidal fringe data with a low-pass filter. However, this is not always satisfactory and information is always lost in filtering operations.

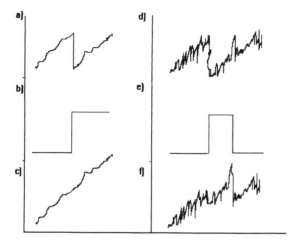

Fig. 11.14. *Effect of additive noise in the wrapped phase data: (a) added noise with a maximum amplitude less than π – the 2π phase discontinuity is correctly located (b) and the data unwrapped; (c) high amplitude noise added to the same data (d) can result in false detection of a 2π phase jump. (e) The resulting unwrapped data (f) retains an erroneous 2π phase jump. (From Robinson, D. W. and Reid, G. T. (eds) (1993) Interferogram Analysis. Digital Fringe Pattern Measurement Techniques, Institute of Physics Publishing. Reproduced by permission of Institute of Physics Publishing and by courtesy of D. W. Robinson)*

For simple unwrapping methods to work, the data must be continuous across the whole image array (no holes in the data) and extend to the boundaries of the sample window. A phase discontinuity might be caused by a rapid change in the measurement parameter, such as a large height step in a component under test. Such defects might appear as a sudden change in the fringe spacing or as a point where the fringe stops (referred to sometimes as a fringe break). In these circumstances, errors in the phase unwrapping are propagated from the defect or hole in the data through the rest of the data array. This is particularly serious if the data is being scanned line by line in one direction. The problems presented by defects or holes in the data becomes even more complex when the shape of the hole becomes complicated. It is to find methods for automatically detecting and compensating for these problems that sophisticated phase-unwrapping algorithms have been developed.

When the terms C and D in Equation (11.62) are both below a certain threshold (i.e. tend to zero) the result of the arctangent calculation becomes indeterminate. Pixels with such values can be flagged or masked in the computer as being invalid data and omitted from phase-unwrapping calculations. The masked pixels can then be replaced by a local average of valid pixels by a final smoothing operation performed on the final unwrapped phase data.

11.5.2 Phase-unwrapping techniques

The basic principle of phase-unwrapping is to 'integrate' the wrapped phase ϕ (in units of 2π) along a path through the data. At each pixel the phase gradient is calculated by differentiation:

$$\Delta\phi = \phi_n - \phi_{n-1} \qquad (11.63)$$

where n is the pixel number. If $|\Delta\phi|$ exceeds a certain threshold such as π, then a phase fringe edge (2π discontinuity) is assumed. This phase jump is corrected by the addition or subtraction of 2π according to the sign of $\Delta\phi$.

Although phase-unwrapping has been performed by analogue circuits as part of an electronic phase measurement process (Mertz 1983), most workers use digital processing techniques to apply this unwrapping process as a logical sequence of program steps. The most common principle used to correct for missed 2π phase jumps is based on the fact that the phase difference between any two points measured by integrating phase along a path between the two points is independent of the route chosen as long as the route does not pass through a phase discontinuity. Thus phase-unwrapping methods may be divided into path-dependent methods and path-independent methods.

11.5.3 Path-dependent methods

The simplest of the phase-unwrapping methods involves a sequential scan through the data, line by line, see Figure 11.15. At the end of each line, the phase difference between the last pixel and the pixel on the line below is determined and the line below is scanned in the reverse direction. In other words a two-dimensional data array is treated like a folded one-dimensional data set. This approach is successful when applied to high-quality data, but more complex variations are necessary in the presence of noise. These include multiple scan directions (Robinson and Williams 1986), spiral scanning (Vrooman and Maas 1989) and counting around defects (Huntley 1989).

Fig. 11.15. *Line by line sequential scanning path*

One of the most popular approaches that is often used to avoid phase errors propagating through the data array is to unwrap the regions of 'good' pixel data first. The 'bad' pixels with high measurement uncertainty are then unwrapped but data propagation errors are then confined to small regions. This is referred to as pixel queuing. A common method is to place the pixel address on a processing queue so that the pixels having the smallest phase difference between neighbouring pixels are processed first (Schorner *et al.* 1991). This approach means that regions of small phase gradients and low noise data are unwrapped first.

Another method of defining good pixels is to examine the interference fringe contrast (visibility) at the pixel under consideration. It is important with these techniques that the order of pixel processing is such that the unwrapping process propagates along connected paths through the array. This can be achieved in different ways (Schorner *et al.* 1991; Towers *et al.* 1991).

Another variation of path-dependent methods is to segment the image into regions containing no phase ambiguities (Kwon *et al.* 1987; Gierloff 1987) or to divide the data array into square tiles or sub-arrays (Towers *et al.* 1991; Stephenson *et al.* 1994). Then the phase data at the edges of adjacent regions or tiles are compared and arranged based on the difference value that most edge pixels agree on.

11.5.4 Path-independent methods

A path-independent method due to Ghiglia *et al.* (1987) (Ghiglia and Romero 1994) is performed in the following way: In a 3×3 mask, the phase difference between the phase p of the central pixel and its four nearest neighbours in the horizontal and vertical directions are calculated. If one of the differences is greater than π in absolute value, $+2\pi$ or -2π is added to p dependent on the majority of the four differences being positive or negative. When there are two positive and two negative differences, an arbitrary decision is taken to add 2π. When none of the absolute differences exceed π, then p remains unchanged. Figure 11.16(a, b) shows an array of wrapped phase data before and after application of one iteration of this algorithm. After several repetitions (iterations) a checker board pattern is seen to spread across the entire array (Figure 11.16(c). One then comes to a point where further repetitions do not change the array further but where every second iteration results in an identical array. At this stage a global iteration is performed by replacing each pixel value by the value of the mean of each pair of pixels in the current and the preceding array. After application of this step, the new phase array resembles the starting array (of wrapped data) except that one less fringe is present (Figure 11.16(d)). The effect has been to unwrap one fringe. Repeated application of this process (local iteration followed by global iteration) removes one fringe at a time until the entire array is unwrapped (Figure 11.16(e)).

As can be easily realized, this algorithm is processing intensive. It is, however, very immune to noise and artifacts.

For more details about phase unwrapping, see Robinson (1993); Judge and Bryanston-Cross (1993).

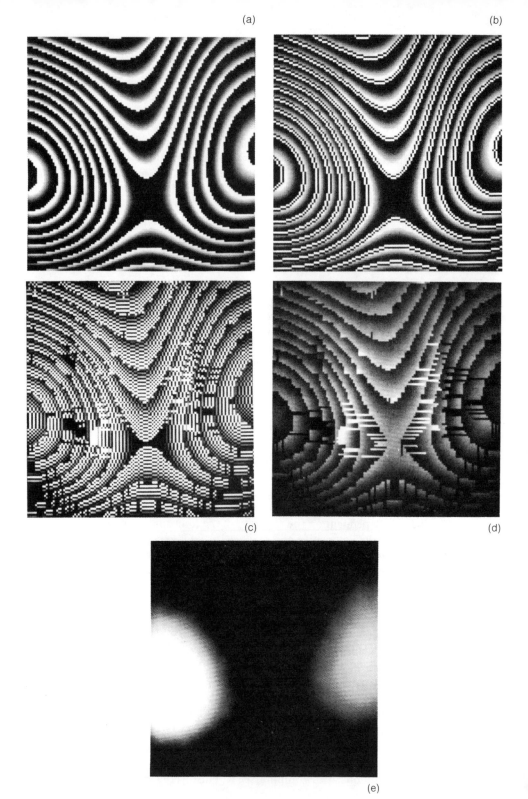

Fig. 11.16. *A series of iterations of the automation-unwrapping algorithm; (a) before iteration; (b) after the first local iteration; (c) after ten local iterations, (d) the first global iteration (averaged state); (e) complete phase unwrapping after 710 total local iterations and 13 total global iterations. (From Ghiglia, et al. 1987. Reproduced by permission of The Optical Society of America and by courtesy of D. C. Ghiglia)*

PROBLEMS

11.1 Suppose we have measured the intensities f_1, f_2, f_4 and f_5 at the pixels $x_1 = -2$, $x_2 = -1$, $x_3 = 0$, $x_4 = 1$ and $x_5 = 2$ respectively. We want to find the fringe maximum by fitting these values to a quadratic curve. Use the formula developed in Appendix C to find an expression for the position x_p of the fringe maximum.

11.2 We want to compare the accuracy when fitting five points and when fitting three points to a quadratic curve. Assume that the intensity distribution along the pixel position x, ideally is given by

$$I(x) = 100\left[1 + \cos\left(\frac{2\pi}{5}x\right)\right]$$

i.e. the fringe period is five pixels with a maximum at $x = 0$.

Now assume that all the measured grey level values follows this distribution (to the nearest integer grey value) except the value measured at $x = 1$, which deviates by 20 percent from the ideal value.

Find the measurement error in pixels when using three-point and five-point curve fitting.

11.3 The phase in a three-frame technique is given by

$$f = \tan \phi = \left(\frac{I_2 - I_3}{I_2 - I_1}\right)$$

Consider I_1, I_2 and I_3 as three mutually independent and normally distributed variables with the same standard deviation σ_1.
(a) Apply the usual law of error estimation to calculate the relative error σ_f/f.
(b) Find an expression for $g = \sin \phi$ and calculate in the same way σ_g/g. What quantity seems to be less sensitive to errors in I_1, I_2 and I_3?

11.4 Consider a four-frame technique with the following phase steps: $\alpha_i = -3\pi/4, -\pi/4, \pi/4, 3\pi/4$. Compute the matrices A, A^{-1} and B. Compute ϕ.

12 Some other optical metrology techniques

In this chapter we briefly mention some other optical metrology techniques.

12.1 LASER DOPPLER VELOCIMETRY (LDV)

As the name (also termed laser Doppler anemometry (Durst *et al.* 1991), LDA) indicates, this is a method for measuring the velocity of, for example, moving objects or particles. It is based on the Doppler effect, which explains the fact that light changes its frequency (wavelength) when detected by a stationary observer after being scattered from a moving object.

This is in analogy with the classical example for acoustical waves when the whistle from a train changes from a high to a low tone as the train passes by.

Here we give an alternative description of the method. Consider Figure 12.1 where a particle is moving in a test volume where two plane waves are interfering at an angle α. In Section 3.3 it was found that these two waves will form interference planes which are parallel to the bisector of α and separated by a distance equal to (cf. Equation (3.21))

$$d = \frac{\lambda}{2 \sin \alpha/2} \tag{12.1}$$

As the particle moves through the test volume, it will scatter light when it is passing a bright interference fringe and scatter no light when it is passing a dark interference fringe. The resulting light pulses can be recorded by a detector placed as in Figure 12.1.

For a particle moving in the direction normal to the interference planes with a velocity v, the time lapse between successive light pulses becomes

$$t_D = \frac{d}{v} \tag{12.2}$$

and thus the frequency

$$f_D = 1/t_D = \frac{2v \sin \alpha/2}{\lambda} \tag{12.3}$$

If there are many particles of different velocities, one will get many different frequencies. They can be recorded on a frequency analyser and the resulting frequency spectrum will tell how the particles are distributed among the different velocities.

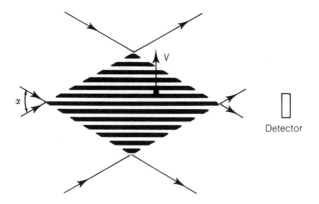

Fig. 12.1. *Laser Doppler velocimetry*

This method does not distinguish between particles moving in opposite directions. If the direction of movement is unknown, one can modulate the phase of one of the plane waves (by means of, for example, an acousto-optic modulator) thereby making the interference planes move parallel to themselves with a known velocity. This velocity will then be subtracted when the particles are moving in the same direction and added when moving in the opposite direction.

In Figure 12.1, the particles pass between the light source and the detector. If the particles scatter enough light, the detector can also be placed on the same side of the test volume as the light source (the laser). Many other configurations of the light source and the detector are described in the literature. For example, one of the two waves can be directly incident on the detector, or it is possible to have one single wave and many detectors.

Laser Doppler velocimetry can be applied for measurement of the velocity of moving surfaces, turbulence in liquids and gases, etc. In the latter cases, the liquid or gas must be seeded with particles. Examples are measurements of stream velocities around ship propellers, velocity distributions of oil drops in combustion and diesel engines, etc.

12.2 ELLIPSOMETRY

Ellipsometry (Passaglia 1964; Neal and Fane 1973) is a method for the measurement of optical constants (refractive index) of surfaces and for measuring the thickness of thin layers. It is based on the fact that light generally changes its state of polarization upon reflection; see Section 9.4.

Since the two reflection coefficients are not equal, the light reflected from a surface will generally become elliptically polarized, hence the name ellipsometry. A fraction of the light, determined by the transmittance coefficients, will propagate into the second medium. When the surface consists of a thin layer, by taking account of the reflection and transmittance coefficients and of multiple reflections in the layer, the state of polarization of the reflected light will depend both on the optical constants and the thickness of the layer.

Fig. 12.2. *Principle of an ellipsometer. P = polarizer A = analyser*

Figure 12.2 shows an ellipsometer with a polarizer P in the incident and an analyser A in the reflected beam. P and A can be combinations of linear polarizers and retarders. The measurement principle is analogous to the one applied in photoelasticity (see Section 9.6). We shall not discuss ellipsometry in more details here, but mention that layer thickness down to the angstrom (10^{-10} m) range can be measured.

12.3 SURFACE ROUGHNESS MEASUREMENT TECHNIQUES

Light scattered from a rough surface (Beckmann and Spizzichino 1963) can be characterized by many different parameters, such as angular distribution, state of polarization, speckle contrast, etc. Much effort has gone into establishing a one-to-one correspondence between many of these scattering parameters and the surface roughness (Persson 1992; Leonhardt *et al.* 1988; Schiffer 1987) to be applied as the principle in a non-contact method for surface roughness measurement. So far, however, no-one has succeeded in making a reliable instrument for this purpose based on such principles. The reason may to a great extent be that the characterization of a rough surface is in itself a complex task.

The best-established light-scattering method is the so-called gloss measurement technique. It consists essentially of measuring the ratio between the intensity of the light scattered in the specular direction and the light scattered in the other directions. A drawback of the method is that the scattered intensity depends on the wavelength of the light and the pigmentation of the surface.

Other promising methods are speckle contrast measurements (Pedersen 1976) and off-specular peaks (Gåsvik 1980, 1981) of polarized light scattered from a rough surface. For smooth surfaces, the roughness ($< \lambda$) can be measured interferometrically.

12.4 FIBRE OPTICS

12.4.1 Light propagation through optical fibres

More extensive treatments on optical fibres can be found in Senior (1985), Palais (1992), Keiser (1991) and Yu and Khoo (1990).

Figure 12.3 shows the basic construction of an optical fibre. It consists of a central cylindrical core with refractive index n_1, surrounded by a layer of material called the cladding with a lower refractive index n_2. In the figure a light ray is incident at the end of the fibre at an angle θ_0 to the fibre axis. This ray is refracted at an angle θ_1 and incident at

Cone of
acceptance

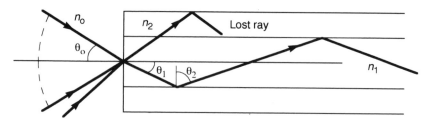

Fig. 12.3. *Basic construction of an optical fibre*

the interface between the core and the cladding at an angle θ_2. From Snell's law of refraction we have

$$n_0 \sin \theta_0 = n_1 \sin \theta_1 \tag{12.4}$$

where n_0 is the refractive index of the surrounding medium. From the figure, we see that

$$\theta_1 = \frac{\pi}{2} - \theta_2 \tag{12.5}$$

If θ_2 is equal to the critical angle of incidence (cf. Section 9.4), we have

$$\sin \theta_2 = \frac{n_2}{n_1} \tag{12.6}$$

which combined with Equations (12.4) and (12.5) gives

$$\theta_0 \equiv \theta_a = \sin^{-1} \left[\frac{\sqrt{n_1^2 - n_2^2}}{n_0} \right] \tag{12.7}$$

For $\theta_0 < \theta_a$ the light will undergo total internal reflection at the interface between the core and the cladding and propagate along the fibre by multiple reflections at the interface, ideally with no loss. For $\theta_0 > \theta_a$ some of the light will transmit into the cladding and after a few reflections, most of the light will be lost.

This is the principle of light transmission through an optical fibre. The angle θ_a is an important parameter when coupling of the light into a fibre, usually given by its numerical aperture NA:

$$NA = n_0 \sin \theta_a = \sqrt{n_1^2 - n_2^2} \tag{12.8}$$

In practice, coupling of the light into the fiber can be accomplished with the help of a lens, see Figure 12.4(a) or by putting the fibre in close proximity to the light source and linking them with an index-matching liquid to reduce reflection losses, Figure 12.4(b). When using the method in Figure 12.4(a), it is important to have the angle of the incident cone less than θ_a to get maximum coupling efficiency.

The above description of light propagation through an optical fibre is not fully complete. To gain better understanding, the fibre must be treated as a waveguide and the

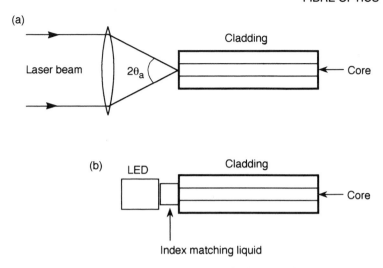

Fig. 12.4. *Coupling of light into a fibre by means of (a) a lens, (b) index-matching liquid*

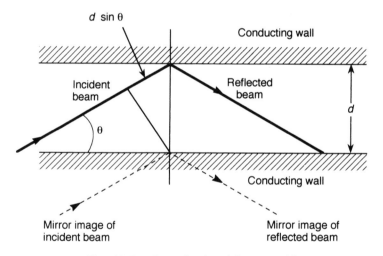

Fig. 12.5. *A conducting slab waveguide*

electromagnetic nature of the light must be taken into account. If a waveguide consisting of a transparent layer between two conducting walls is considered, the electric field across the waveguide will consist of interference patterns between the incident and reflected fields, or equivalently, between the incident field and its mirror image, see Figure 12.5. The path-length difference Δl between these fields is seen from the figure to be

$$\Delta l = d \sin \theta \tag{12.9}$$

where d is the waveguide diameter and θ is the angle of the incident beam. From the boundary conditions for such a waveguide we must have destructive interference at the walls, i.e. the path-length difference must be equal to an integral number of half the wavelength:

$$\Delta l = m\frac{\lambda}{2} \tag{12.10}$$

which gives

$$\sin\theta = \frac{m\lambda}{2d} \tag{12.11}$$

where m is an integer. Thus we see that only certain values of the angle of incidence are allowed. Each of the allowed beam directions is said to correspond to different modes of wave propagation in the waveguide. The field distribution across the waveguide for the lowest-order guided modes in a planar dielectric slab waveguide are shown in Figure 12.6. This guide is composed of a dielectric core (or slab) sandwiched between dielectric claddings of lower refractive index. As can be seen, the field is non-zero inside the cladding. This is not in contradiction with the theory of total internal reflection (see Section 9.4) which predicts an evanescent wave decaying very rapidly in the cladding material.

The lowest number of modes propagating through the waveguide occurs when the angle of incidence is equal to θ_a. Then (assuming $n_0 = 1$ for air)

$$\sin\theta_a = \frac{m\lambda}{2d} = \sqrt{n_1^2 - n_2^2} \tag{12.12}$$

or

$$\frac{d}{\lambda} = \frac{m}{2\sqrt{n_1^2 - n_2^2}} \tag{12.13}$$

To have only the lowest-order mode ($m = 0$) propagating through the waveguide, we therefore must have

$$\frac{d}{\lambda} < \frac{1}{2\sqrt{n_1^2 - n_2^2}} \tag{12.14}$$

An exact waveguide theory applied to an optical fibre is quite complicated, but the results are quite similar. The condition for propagating only the lowest-order mode in an optical

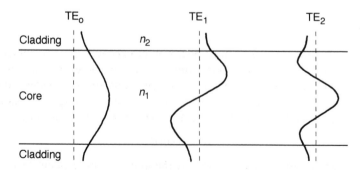

Fig. 12.6. *Electric field distribution of the lowest-order guided transversal modes in a dielectric slab waveguide*

fibre then becomes

$$\frac{d}{\lambda} < \frac{2.405}{2\pi\sqrt{n_1^2 - n_2^2}} = \frac{2.405}{2\pi(NA)} = \frac{0.383}{NA} \tag{12.15}$$

A fibre allowing only the lowest-order mode to propagate is called a single-mode fibre, in contrast to a multimode fibre which allows several propagating modes.

12.4.2 *Attenuation and dispersion*

That light will propagate through a fibre by multiple total internal reflections without loss is an idealization. In reality the light will be attenuated. The main contributions to attenuation is scattering (proportional to λ^{-4}) in the ultraviolet end of the spectrum and absorption in the infra-red end of the spectrum. Therefore it is only a limited part of the electromagnetic spectrum where fibre systems are practical. Figure 12.7 shows the attenuation as a function of wavelength for silica glass fibres. Here are also shown the three major wavelength regions at which fibre systems are practical. These regions are dictated by the attenuation, but also by the light sources available.

Another source of loss in fibre communication systems is dispersion. Dispersion is due to the fact that the refractive index is not constant, but depends on the wavelength, i.e. $n = n(\lambda)$. In fibre systems one talks about material dispersion and waveguide dispersion. Here we will briefly mention material dispersion. That the refractive index varies with wavelength means that a light pulse from a source of finite spectral width will broaden as it propagates through the fibre due to the different velocities for the different wavelengths. This effect has significant influence on the information capacity of the fibre. The parameter describing this effect is the pulse spread per unit length denoted $\Delta\tau/L$ where $\Delta\tau$ is the

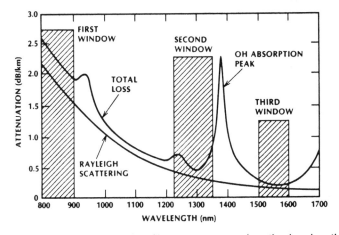

Fig. 12.7. *Attenuation in a silica glass fibre versus wavelength showing the three major wavelength regions at which fibre systems are most practical. (From Palais, J. C. (1992) Fiber Optic Communications (3rd edn), Prentice Hall, Englewood Cliffs, N.J.) Reproduced by permission of Prentice Hall Inc.)*

difference in travel time for two extreme wavelengths of the source's spectral distribution through the length L. This gives

$$\frac{\Delta\tau}{L} = \Delta\left(\frac{1}{v_g}\right) \tag{12.16}$$

In dispersive media a light pulse propagates at the group velocity (Senior 1985) defined by

$$v_g = \frac{d\omega}{d\beta} \tag{12.17}$$

With the relations

$$\omega = kc = \frac{2\pi c}{\lambda} \tag{12.18a}$$

$$\beta = kn = \frac{2\pi n}{\lambda} \tag{12.18b}$$

we get

$$\frac{1}{v_g} = \frac{d\beta}{d\omega} = \frac{d\beta}{d\lambda}\frac{d\lambda}{d\omega} = \left(\frac{-\lambda^2}{2\pi c}\right)2\pi\left(\frac{1}{\lambda}\frac{dn}{d\lambda} - \frac{n}{\lambda^2}\right) = \frac{1}{c}\left(n - \lambda\frac{dn}{d\lambda}\right) \tag{12.19}$$

This gives

$$\frac{\Delta\tau}{L} = \Delta\left(\frac{1}{v_g}\right) = \Delta\left(\frac{n - \lambda\,dn/d\lambda}{c}\right) \tag{12.20}$$

The pulse spread per unit length per wavelength interval $\Delta\lambda$ becomes

$$\frac{\Delta\tau}{L\Delta\lambda} = \frac{d}{d\lambda}\left(\frac{1}{v_g}\right) = \frac{d}{d\lambda}\left(\frac{n}{c} - \frac{\lambda\,dn}{c\,d\lambda}\right) = -\frac{\lambda\,d^2n}{c\,d\lambda^2} \tag{12.21}$$

The material dispersion is defined as $M = (\lambda/c)(d^2n/d\lambda^2)$. The pulse spread per unit length then can be written as

$$\frac{\Delta\tau}{L} = -M\,\Delta\lambda \tag{12.22}$$

The refractive index for pure silicon dioxide (SiO_2) glass used in optic fibres has the wavelength dependence shown in Figure 12.8(a). At a particular wavelength λ_0, there is an inflection point on the curve. Because of this, $d^2n/d\lambda^2 = 0$ at λ_0 as seen from the curve of the second derivative in Figure 12.8(b). For pure silica, the refractive index is close to 1.45 and the inflection point is near 1.3 μm. Therefore this wavelength is very suitable for long distance optical fibre communication.

12.4.3 Different types of fibres

Another construction than the step-index (SI) fibre sketched in Figure 12.3 is the so-called graded-index (GRIN) fibre. It has a core material whose refractive index varies with distance from the fibre axis. This structure is illustrated in Figure 12.9. As should be easily realized, the light rays will bend gradually and travel through a GRIN fibre in the

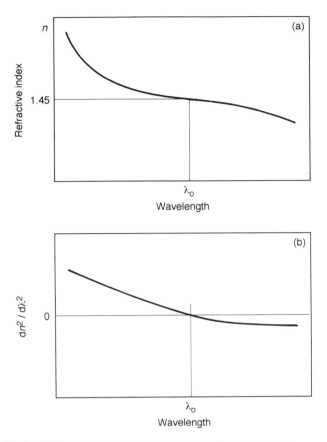

Fig. 12.8. *(a) Refractive index versus wavelength for SiO$_2$ glass. (b) The second derivative of the curve in (a)*

oscillatory fashion sketched in Figure 12.9(d). As opposed to an SI fibre, the numerical aperture of a GRIN fibre decrease with radial distance from the axis. For this reason, the coupling efficiency is generally higher for SI fibres than for GRIN fibres, when each has the same core size and the same fractional refractive index change.

Conventionally, the size of a fibre is denoted by writing its core diameter and then its cladding diameter (both in micrometers) with a slash between them. Typical dimensions of SI fibres are 50/125, 100/140 and 200/230 and typical dimensions of multimode GRIN fibres are 50/125, 62.5/125 and 85/125.

SI fibres have three common forms: (1) a glass core cladded with glass, (2) a silica glass core cladded with plastic (termed plastic-cladded silica (PCS) fibres), and (3) a plastic core cladded with another plastic. All-glass fibres have the lowest losses and the smallest pulse spreading, but also the smallest numerical aperture. PCS fibres have higher losses and larger pulse spreads and are suitable for shorter links, normally less than a few hundred metres. Their higher NA increase the coupling efficiency. All-plastic fibres are used for path lengths less than a few tens of meters. Their high NA gives high coupling efficiency. Single-mode fibres have the highest information capacity. GRIN fibres can transmit at higher information rates than SI fibres. Table 12.1 shows representative numerical values

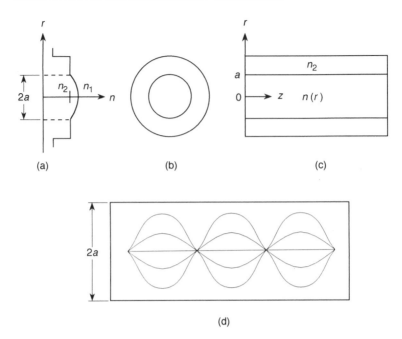

(a)   (b)   (c)

(d)

Fig. 12.9. *Graded index fibre: (a) refractive index profile; (b) end view; (c) cross-sectional view; (d) ray paths along a GRIN fibre*

Table 12.1 (From Palais, J. C. (1992) *Fiber Optic Communication* (3rd edn), Prentice Hall, Englewood Cliffs, New Jersey). Reproduced by permission

Description	Core Diameter (μm)	NA	Loss (dB/km)	$\Delta(\tau/L)$ (ns/km)	Source	Wavelength (nm)
Multimode						
Glass						
SI	50	0.24	5	15	LED	850
GRIN	50	0.24	5	1	LD	850
GRIN	50	0.20	1	0.5	LED, LD	1300
PCS						
SI	200	0.41	8	50	LED	800
Plastic						
SI	1000	0.48	200	—	LED	580
Single mode						
Glass	5	0.10	4	<0.5	LD	850
Glass	10	0.10	0.5	0.006	LD	1300
Glass	10	0.10	0.2	0.006	LD	1550

Polyurethane, 3.8 mm

Kevlar, 2 mm

Hytrel secondary buffer, 1 mm

Silastic primary buffer, 0.4 mm

Fibre, 0.23 mm

Fig. 12.10. *Light-duty, tight-buffer fibre cable (Siecor Corporation). The dimensions given are the diameters. (From Palais, J. C. (1992) Fiber Optic Communications (3rd edn), Prentice Hall, Englewood Cliffs, N.J.) Reproduced by permission of Prentice Hall*

of important properties for the various fibres. Somewhat different characteristics may be found when searching the manufacturers' literature.

The amount of protection against the environment of a fibre varies from one application to another. Various cable designs have been implemented. A representative light-duty cable is sketched in Figure 12.10. This cable weighs 12.5 kg/km and can withstand a tensile load of 400 N during installation and can be loaded up to 50 N in operation.

Fibre-optic communications developed very quickly after the first low-loss fibres were produced in 1970. Today, over 10 million km of fibre have been installed worldwide, numerous submarine fibre cables covering the Atlantic and Pacific oceans and many other smaller seas are operational. In addition, installation of fibre-optic local area networks (LANs) is increasing.

12.4.4 Fibre-optic sensors

Over the past few years, a significant number of sensors using optical fibres have been developed (Kyuma *et al.* 1982; Culshaw 1986; Udd (1991, 1993). They have the potential for sensing a variety of physical variables, such as acoustic pressure, magnetic fields, temperature, acceleration and rate of rotation. Also sensors for measuring current and voltage based on polarization rotation induced by the magnetic field around conductors due to the Faraday effect in optical fibres have been developed. It should also be mentioned that a lot of standard optical equipment has been redesigned using optical fibres. The Laser Doppler velocimeter is an example where optical fibres have been incorporated to increase the versatility of the instrument.

Figure 12.11 shows some typical examples of fibre-optic sensors. In Figure 12.11(a) a thin semiconductor chip is sandwiched between two ends of fibres inside a steel pipe. The light is coming through the fibre from the left and is partly absorbed by the semiconductor. This absorption is temperature-dependent and the amount of light detected at the end of the fibre to the right is therefore proportional to the temperature and we have a fibre-optic temperature sensor. Figure 12.11(b) shows a simplified sketch of a pressure-sensing system. The optical fibre is placed between two corrugated plates. When pressure is

Fig. 12.11. (a) Fibre-optic temperature sensor. (b) Fibre-optic pressure sensor

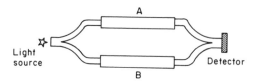

Fig. 12.12. Interferometric fibre-optic sensor

applied to the plates, the light intensity transmitted by the fibre changes, owing to microbending loss. Such systems have also been applied as hydrophones and accelerometers.

Figure 12.12 shows the principle of a class of fibre-optic sensors based on interferometry. The fibres A and B can be regarded as either arm in a Mach–Zehnder interferometer. The detector will record an intensity which is dependent on the optical path-length difference through A and B. When, for example, fibre A is exposed to loads such as tension, pressure, temperature, acoustical waves, etc., the optical path length of A will change and one gets a signal from the detector varying as the external load.

Figure 12.13 shows a special application of optical fibres. In Figure 12.13(a) two fibre bundles, A and B, are mixed together in a bundle C in such a way that every second fibre in the cross-section of C comes from, say, bundle A. Figure 12.13(b) shows two neighbouring fibres, A and B. Fibre A emits a conical light beam. Fibre B will receive light inside a cone of the same magnitude. If a plane surface is placed a distance *l* in front of the fibre ends, light will be scattered back and the amount of light received by fibre B will be proportional

Fig. 12.13

to the area of overlap between the two cones as illustrated in the figure. A curve describing the relation between the received light intensity I_B versus the distance l therefore will look like that given in Figure 12.13(c). For a distance l so long that the whole cross-section of fibre B is covered with light, I_B will have its maximum. A further increase in l will decrease the value of I_B.

The fibre bundles coupled together as in Figure 12.13(a) thus can be used as a non-contact distance sensor (Cook and Hamm 1979; Philips 1980). A light source emits light into bundle A and a detector is placed at the end of bundle B. By placing the end of bundle C close to a surface, the detector will give a signal which is proportional to the distance from the fibre end to the surface, as long as one is working within the linear portion of the left flank of the curve in Figure 12.13(c). The sensitivity limit of this type of sensor can be down to nanometres and it is especially well-suited to the measurement of small vibration amplitudes. By mounting a membrane in front of fibre bundle C, this equipment can be used for pressure measurements. In this way it has been applied for the measurement of blood pressure and electric fields inside big power cables.

One of the most important and potentially low-cost advantages of fibre-optic sensors is their multiplexing capability (Berthold 1993). For example, multiple fibre-optic accelerometers can be linked to a common transmission 'bus' fibre. These accelerometers can all be remotely interrogated in turn from a central processor without any intervening electronics. Either wavelength division (WDM) or time division (TDM) multiplexing methods could be employed. In contrast, commercially available piezo-electric accelerometers each requires its own charge amplifier/signal conditioner to achieve low-noise operation, and thus these devices are not compatible with a low-cost multiplexing architecture. A system based on wavelength division multiplexing is sketched in Figure 12.14. Here the different wavelengths from the light source (an LED) are directed to each accelerometer by the WDM units. At the receiving signal processing unit, the intensity of each separate wavelength is analysed.

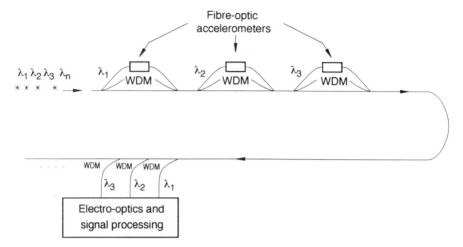

Fig. 12.14. *Multiplexing of fibre-optic sensors (accelerometers) using WDM*

12.5 LASER UP-LINING. RANGING

Finally we shall mention the most immediate application of the laser beam, namely as a ruler. This type of applications is quite common within building and construction work. In tunnel drilling, the drilling machine is steered so as to keep a laser beam falling onto a reference point on the machine. Similar principles are applied in the construction of pipelines and railways. Equipment, specially designed for such purposes is today commerically available. Also available are laser plumbs that automatically emit the beam in the horizontal direction. By means of mirrors, the beam can be rotated in the horizontal plane. This can be used for floor alignment.

The reading of the position of the laser beam can be done visually, but better accuracy is obtained by means of detectors. A special type of position-sensing detector gives an output voltage proportional to the distance of the light spot from a reference point on the detector. It is the centroid of the light spot that is sensed and thus the position measurement is independent of the spot diameter as long as it is inside the detector area.

The position of an unexpanded laser beam directly incident on such a detector can be determined to an accuracy of less than 1 µm. Also photodiodes placed in a matrix can be used for detecting the location of a light spot.

Position-sensing detectors combined with lenses, mirrors and prisms can be combined into many different displacement-measuring set-ups, such as those shown in Figure 12.15. The method illustrated in Figure 12.15(d) is especially interesting since it is a non-contact method. Here the laser beam forms a light spot on the test object surface. This spot is imaged, by means of a lens, onto the detector. When the object moves, the image of the light spot moves on the detector area.

Assume that the object moves a distance s normal to its surface. From the figure, using simple trigonometric relations, we find that the corresponding movement of the imaged

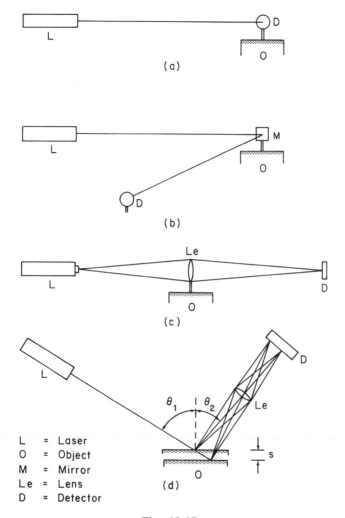

L = Laser
O = Object
M = Mirror
Le = Lens
D = Detector

Fig. 12.15

spot on the detector is given by

$$s' = m \cdot s(\tan \theta_1 \cos \theta_2 + \sin \theta_2) \tag{12.23}$$

Here θ_1 is the angle of incidence of the laser beam, θ_2 is the viewing angle and m is the transversal magnification of the lens. We see that the movement can be magnified by the lens, thereby increasing the sensitivity. However, the size of the light spot will also be magnified, and this must always be inside the detector area to avoid measurement errors, thus limiting the usable magnification.

The laser beam can also be used for the measurement of long distances. One method is to measure the time of flight for a light pulse from the laser to the target and back to a detector. Another method is to modulate a continuous wave. The laser has been used for measuring the distance to the Moon. To increase the amount of reflected light, astronauts have placed retroflective mirrors on the lunar surface.

PROBLEMS

12.1 Consider an optical fibre designed to guide monochromatic light of $\lambda = 400$ nm. The diameter of the core is $10\,\mu$m. The refractive indices of the core and the cladding are $n_1 = 1.5$ and $n_2 = 1.35$. Calculate
(a) the allowable angles of the incident beams;
(b) the largest allowable angle of the incident beam;
(c) the actual acceptance angle of the fibre;
(d) the numerical aperture of the fibre.

12.2 The refractive index of a single-mode fibre core and its cladding are $n_1 = 1.5$ and $n_2 = 1.47$, and the wavelength of the light is $\lambda = 600$ nm. Calculate the core diameter required.

12.3 (a) Find the amount of pulse spreading in pure silica for an LED operating at 820 nm and having a 20 nm spectral width. The path is 10 km long and the material dispersion $M = 110\,\text{ps}/(\text{nm km})$. (1 ps $= 10^{-12}$ s)
(b) Repeat the problem when $\lambda = 1.5\,\mu$m and $\Delta\lambda = 50$ nm. $M = -15\,\text{ps}/(\text{nm km})$.
(c) Repeat (a) and (b) when the source is a laser diode with a 1 nm spectral width.

12.4 Due to the pulse spread $\Delta\tau$, the wavelength components of a sinusoidally modulated beam will have different transit times. When the delay between the fastest and slowest wavelength is equal to half the period T of the modulated signal, the two components are in antiphase and will cancel each other. This sets a limit on the allowable pulse spread given by

$$\Delta\tau \leqslant \frac{T}{2}$$

which gives for the modulation frequency

$$f = \frac{1}{T} \leqslant \frac{1}{2\Delta\tau}$$

Calculate the frequency limit for a 10 km fibre link for the examples given in Problem 12.3.

12.5 The number of modes in a step-index fibre is given by

$$N = 2\left(\frac{\pi a}{\lambda}\right)^2 (n_1^2 - n_2^2)$$

where a is the core radius and λ is the free-space wavelength.
(a) Compute the number of modes for a fibre whose core diameter is $50\,\mu$m. Assume that $n_1 = 1.48$ and $n_2 = 1.46$. Let $\lambda = 820$ nm.
(b) Consider a SI fibre with $n_1 = 1.5$ and $n_2 = 1.485$ at 820 nm. If the core radius is $50\,\mu$m, how many modes can propagate.
(c) Repeat problem (b) if the wavelength is changed to $1.2\,\mu$m.
(d) What is the maximum core radius allowed for a glass fibre having $n_1 = 1.465$ and $n_2 = 1.46$ if the waveguide is to support only one mode at a wavelength of 1250 nm?

12.6 Consider the non-contact fibre-optic sensor illustrated in Figure 12.13. Assume that the intensity received by fibre B is proportional to $1/l^4$. Given that the area of overlap A between two circles of radius r and centre separation a is

$$A = 2a \sqrt{r^2 - \left(\frac{a}{2}\right)^2}$$

(a) Show that the received intensity I_B is indeed like that sketched in Figure 12.13(c). Let the distance at which the cones from fibre A and B start to overlap be l_0.
(b) Find the distance l_m where I_B is maximum in terms of l_0.

Appendix A   Complex numbers

The extension of the notion of numbers from real to complex numbers consists of changing their representation from points on a line to points in a plane.

In Figure A.1, the complex number z is represented by a point with coordinates x and y along the real and imaginary coordinate axes. The unit along the imaginary axis is equal to $i = \sqrt{-1}$ such that z can be written as

$$z = x + iy \tag{A.1}$$

where

$$x = \operatorname{Re}\{z\} = \text{the real part of } z$$

$$y = \operatorname{Im}\{z\} = \text{the imaginary part of } z$$

In polar coordinates (see Figure A.1), Equation (A.1) becomes

$$z = r(\cos\phi + i\sin\phi) \tag{A.2}$$

where $r = |z| =$ the absolute value (modulus) or the length of z. We see that

$$r = \sqrt{x^2 + y^2} \tag{A.3}$$

$$\tan\phi = y/x \tag{A.4}$$

By means of Euler's formula

$$e^{i\phi} = \cos\phi + i\sin\phi \tag{A.5}$$

z can be written as

$$z = re^{i\phi} \tag{A.6}$$

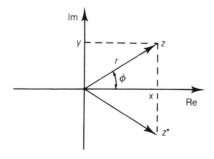

Fig. A.1

Those who are not familiar with the Euler formula can convince themselves by taking the differential

$$dz = r(-\sin\phi + i\cos\phi)d\phi$$
$$= ir(\cos\phi + i\sin\phi)d\phi = iz\,d\phi \qquad (A.6a)$$

and then integrate

$$\int dz/z = i\int d\phi \qquad (A.6b)$$

and the Euler formula is obtained.

The complex conjugate z^* of z is found by changing i into $-$i everywhere it occurs:

$$z^* = x - iy = r(\cos\phi - i\sin\phi) = re^{-i\phi} \qquad (A.7)$$

The geometrical interpretation of complex conjugation is to find the mirror image of z about the real axis (see Figure A.1).

Addition (subtraction) of two complex numbers is done by simply adding (subtracting) the real and imaginary parts separately, viz.

$$z_1 \pm z_2 = (x_1 + iy_1) \pm (x_2 + iy_2) = (x_1 \pm x_2) + i(y_1 \pm y_2) \qquad (A.8)$$

Multiplication and division are most easily performed using polar notation

$$z_1 z_2 = r_1 r_2 e^{i(\phi_1 + \phi_2)} \qquad (A.9)$$

$$z_1 : z_2 = \frac{r_1}{r_2} e^{i(\phi_1 - \phi_2)} \qquad (A.10)$$

From this, we find that

$$|z| = \sqrt{zz^*} = r \qquad (A.11)$$
$$|z|^2 = zz^* = r^2 \qquad (A.12)$$

For

$$z = z_1 + z_2 \qquad (A1.13)$$

we get

$$|z|^2 = |z_1 + z_2|^2 = (z_1 + z_2)(z_1^* + z_2^*)$$
$$= |z_1|^2 + |z_2|^2 + z_1 z_2^* + z_1^* z_2$$
$$= r_1^2 + r_2^2 + 2r_1 r_2 \cos(\phi_1 - \phi_2) \qquad (A.14)$$

Appendix B Fourier optics

B.1 THE FOURIER TRANSFORM

The two-dimensional Fourier transform of the function $g(x, y)$ is defined as

$$G(f_x, f_y) = \mathcal{F}\{g(x, y)\} = \int_{-\infty}^{\infty} \int g(x, y)e^{-i2\pi(f_x x + f_y y)}dx\,dy \tag{B.1}$$

where from the inverse transform is given by

$$g(x, y) = \mathcal{F}^{-1}\{G(f_x, f_y)\} = \int_{-\infty}^{\infty} \int G(f_x, f_y)e^{i2\pi(f_x x + f_y y)}df_x\,df_y \tag{B.2}$$

The following theorems can by proved by inserting into Equation (B.1).
Assume that

$$\mathcal{F}\{g(x, y)\} = G(f_x, f_y), \text{ and } \mathcal{F}\{h(x, y)\} = H(f_x, f_y)$$

Then
1. *Linearity theorem*

$$\mathcal{F}\{\alpha g(x, y) + \beta h(x, y)\} = \alpha\mathcal{F}\{g(x, y)\} + \beta\mathcal{F}\{h(x, y)\} \tag{B.2a}$$

2. *Similarity theorem*

$$\mathcal{F}\{g(ax, by)\} = \frac{1}{|ab|}G\left(\frac{f_x}{a}, \frac{f_y}{b}\right) \tag{B.2b}$$

3. *Shift theorem*

$$\mathcal{F}\{g(x - a, y - b)\} = G(f_x, f_y)e^{-i2\pi(f_x a + f_y b)} \tag{B.2c}$$

4. *Parseval's theorem*

$$\int_{-\infty}^{\infty} \int |g(x, y)|^2 \, dx\,dy = \int_{-\infty}^{\infty} \int |G(f_x, f_y)|^2 \, df_x\,df_y \tag{B.2d}$$

5. *Convolution theorem*

$$\mathcal{F}\{g(x, y) \otimes h(x, y)\} = \mathcal{F}\left\{\int_{-\infty}^{\infty} \int g(\xi, \eta)h(x - \xi, y - \eta)\,d\xi\,d\eta\right\} = G(f_x, f_y)H(f_x, f_y) \tag{B.2e}$$

6. *Autocorrelation theorem*

$$\mathcal{F}\{g(x, y) \odot g(x, y)\} = \mathcal{F}\left\{\int_{-\infty}^{\infty} \int g(\xi, \eta)g^*(\xi - x, \eta - y)\,d\xi\,d\eta\right\} = |G(f_x, f_y)|^2 \tag{B.2f}$$

7. Fourier integral theorem

$$\mathscr{F}\mathscr{F}^{-1}\{g(x, y)\} = \mathscr{F}^{-1}\mathscr{F}\{g(x, y)\} = g(x, y) \tag{B.2g}$$

A function and its transform can also be represented in polar coordinates such that $\mathscr{F}\{g(r, \theta)\} = G(\rho, \phi)$ where

$$
\begin{aligned}
x &= r\cos\theta & f_x &= \rho\cos\phi \\
y &= r\sin\theta & f_y &= \rho\sin\phi
\end{aligned} \tag{B.3}
$$

An important class of functions are those which possess circular symmetry, that means they are independent of θ, such that

$$g(r, \theta) = g_R(r) \tag{B.4}$$

By substituting Equation (B.3) and (B.4) into (B.1) we get

$$G(\rho, \phi) = \int_0^{2\pi} d\theta \int_0^{\infty} r g_R(r) e^{-i2\pi r\rho(\cos\theta\cos\phi + \sin\theta\sin\phi)} \, dr \tag{B.5}$$

$$= \int_0^{\infty} r g_R(r) \, dr \int_0^{2\pi} e^{-2i\pi r\rho\cos(\theta - \phi)} \, d\theta$$

By applying the identity

$$J_0(a) = (1/2\pi) \int_0^{2\pi} e^{-ia\cos(\theta - \phi)} \, d\theta \tag{B.6}$$

where J_0 is the zerorth-order Bessel function, Equation (B.5) becomes

$$G(\rho) = 2\pi \int_0^{\infty} r g_R(r) J_0(2\pi r\rho) \, dr \tag{B.7}$$

Thus the Fourier transform of a circularly symmetric function is itself circularly symmetric. Equation (B.7) is referred to as the zeroth-order Hankel transform or alternatively as the Fourier-Bessel transform. Its inverse is given as

$$g_R(r) = 2\pi \int_0^{\infty} \rho G(\rho) J_0(2\pi r\rho) \, d\rho \tag{B.8}$$

Thus, for circularly symmetric functions there is no difference between the transform and inverse transform operations.

B.2 SOME FUNCTIONS AND THEIR TRANSFORMS

Below we give the definition of some frequently used functions. They are illustrated in Figure B.1 and some transform pairs are listed in Table B.1.

The delta function

$$\delta(x, y) = \begin{cases} \infty & \text{for } x = y = 0 \\ 0 & \text{otherwise} \end{cases} \tag{B.9}$$

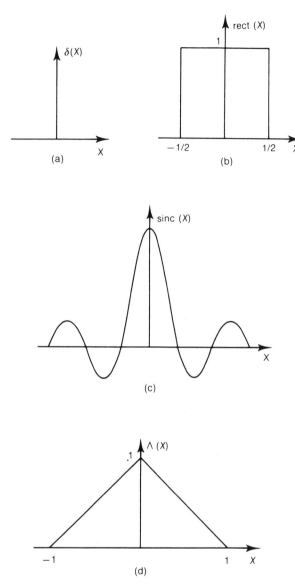

Fig. B.1. *(g) from J. W. Goodman, Introduction to Fourier Optics.* © *1968. Reproduced by permisson of McGraw-Hill Book Company.)*

$\delta(x, y)$ is a discontinuous function used for representing a point source or a focal point. Its integrated area is equal to 1 and it is usually represented by an arrow of height equal to 1. Mathematically, it can be defined as a limiting value. We might also take its transform properties as a definition:

$$\mathscr{F}\{\delta(x, y)\} = 1 \tag{B.10}$$

$$\mathscr{F}\{1\} = \delta(f_x, f_y) \tag{B.11}$$

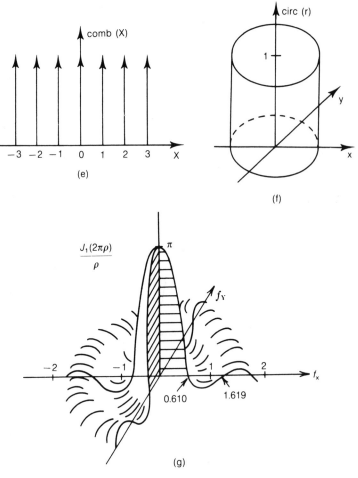

Fig. B.1. *(continued)*

Table B.1. *Transform pairs*

$g(x, y)$	$\mathscr{F}\{g(x, y)\}$
$\delta(x, y)$	1
$\cos(2\pi f_0 x)$	$\frac{1}{2}\delta(f_x - f_0) + \frac{1}{2}\delta(f_x + f_0)$
$\mathrm{rect}(x)\,\mathrm{rect}(y)$	$\mathrm{sinc}(f_x)\,\mathrm{sinc}(f_y)$
$\Lambda(x)\Lambda(y)$	$\mathrm{sinc}^2(f_x)\,\mathrm{sinc}^2(f_y)$
$\mathrm{comb}(x)\,\mathrm{comb}(y)$	$\mathrm{comb}(f_x)\,\mathrm{comb}(f_y)$
$\mathrm{circ}(r)$	$\dfrac{J_1(2\pi\rho)}{\rho}$
$\cos(2\pi f_0 r)$	$-\dfrac{1}{2\pi}f_0(f_0^2 - \rho^2)^{3/2}\,\mathrm{circ}\left(\dfrac{\rho}{f_0}\right)$

$\delta(x, y)$ has the following properties:

$$\delta(ax, by) = \frac{1}{|ab|}\delta(x, y) \tag{B.12a}$$

$$\int_{-\infty}^{\infty}\int \delta(x, y)\,dx\,dy = 1 \tag{B.12b}$$

$$\int\int_{-\infty}^{\infty} g(\xi, \eta)\delta(x - \xi, y - \eta)\,d\xi\,d\eta = g(x, y) \tag{B.12c}$$

Equation (B.12c) is called the shifting property, which is very useful.

Rectangle function

$$\text{rect}(x) = \begin{cases} 1 & \text{for } |x| \leqslant 1/2 \\ 0 & \text{otherwise} \end{cases} \tag{B.13}$$

Sinc function

$$\text{sinc}(x) = \frac{\sin \pi x}{\pi x} \tag{B.14}$$

Triangle function

$$\Lambda(x) = \begin{cases} 1 - |x| & \text{for } |x| \leqslant 1 \\ 0 & \text{otherwise} \end{cases} \tag{B.15}$$

Comb function

$$\text{comb}(ax) = \frac{1}{a}\sum_{n=-\infty}^{\infty} \delta\left(x - \frac{n}{a}\right) \tag{B.16}$$

Circle function

$$\text{circ}((r) = \begin{cases} 1 & \text{for } r \leqslant 1 \\ 0 & \text{otherwise} \end{cases} \tag{B.17}$$

The transform of circ(r) (see Figure (B.1f)) is found by using Equation (B.7) and the identity

$$\int_{0}^{x} \xi J_0(\xi)\,d\xi = x J_1(x) \tag{B.18}$$

where J_1 is the first-order Bessel function. The other transform pairs are easily proved.

B.3 SOME IMPLICATIONS

In Sections 4.4, B.1 and B.2 we have established a powerful tool for the calculation of diffraction patterns and the analysis and synthesis of optical systems. Below, some important implications of the matter given in these sections are discussed. We omit the factor $K/i\lambda z$ in Equation (4.15) and regard the transfer from the diffraction plane to the observation plane as a pure Fourier transform. This we can do, since in calculating the intensity, $1/\lambda^2 z^2$ becomes an unimportant scale factor.

The linearity theorem

Assume that a diffracting screen is given by the transmittance function $g(x, y)$. Another screen, given by the transmittance function $h(x, y) = 1 - g(x, y)$ will be open at places where the first is non-transmitting and vice versa. Two such screens (objects) are called complementary. From the linearity theorem, we find that $\mathscr{F}\{h(x, y)\} = \delta(f_x, f_y) - \mathscr{F}\{g(x, y)\}$. This means that apart from a delta function (a light spot) at the origin, the diffraction patterns from two complementary objects will be equal. This fact is referred to as the Babinet principle and may be useful in cases where it is easier to calculate the diffraction pattern from the complementary object instead of the actual object.

The shift theorem

Consider an object with the transmittance function $g(x, y)$. The function $g(x - a, y - b)$ will represent the same object translated a distance a in the x-direction and a distance b in the y-direction. From the shift theorem we see that apart from a phase factor (which becomes equal to 1 by calculating the intensity), the diffraction pattern is unaffected by a translation in the xy-plane. This fact can be taken advantage of when measuring the dimensions of moving objects by making measurements on the diffraction pattern.

The convolution theorem

The convolution theorem tells that

$$\mathscr{F}\{g(x, y) \otimes h(x, y)\} = G(f_x, f_y)H(f_x, f_y) \tag{B.19}$$

where we have introduced the shorthand notation for the convolution integral:

$$g(x, y) \otimes h(x, y) = \int_{-\infty}^{\infty} \int g(\xi, \eta)h(x - \xi, y - \eta)\,d\xi\,d\eta \tag{B.20}$$

This implies that

$$\mathscr{F}\{g(x, y)h(x, y)\} = G(f_x, f_y) \otimes H(f_x, f_y) \tag{B.21}$$

Geometrically, $g \otimes h$ is given as the area of overlap between g and h as a function of the position of g as g is translated from $-\infty$ to $+\infty$. Thus we have, for example,

$\text{rect}(x) \otimes \text{rect}(x) = \Lambda(x)$. The convolution of a function with the δ-function becomes especially simple since $g(x, y) \otimes \delta(x, y) = g(x, y)$ (cf. Equation (B.12c)), and therefore

$$g(x, y) \otimes \delta(x - a, y - b) = g(x - a, y - b) \tag{B.22}$$

The convolution theorem applied to Equation (B.22) gives

$$\mathcal{F}\{g(x - a, y - b)\} = \mathcal{F}\{g(x, y)\}\mathcal{F}\{\delta(x - a, y - b)\}$$

$$= G(f_x, f_y)e^{-i2\pi(f_a a + f_y b)} \tag{B.23}$$

which is another proof of the fact that the diffraction pattern from an object is unaffected (apart from a phase factor) by translating it in the object plane.

Appendix C   The least squares error method

Consider the equation

$$f = a_0 g^{(0)} + a_1 g^{(1)} + a_2 g^{(2)} + \cdots + a_n g^{(n)} \tag{C.1}$$

where $g^{(i)}$ may be functions of, say, x and a_i are constant coefficients. ($g^{(i)}$ may be functions of $x_1, x_2, \ldots, x_{k-1}$ where $(x_1, \ldots, x_{k-1}, f)$ forms a k-tuple in k-dimensional space). Assume that a_i are unknown coefficients which can be determined by measuring coordinate pairs (x_j, f_j). To solve for the $n+1$ unknown coefficients, we require at least $n+1$ such observations. Such a system is said to be exactly determined. However, the solution of these exact systems is often ill-conditioned (numerically unstable) and it is usually good practice to overdetermine the system by specifying more control points than needed (and hence generate more simultaneous equations). However, an over-determined system does not have an exact solution and there are going to be some errors for some points. The idea, then, is to minimize these errors. A common approach is to minimize the sum of the square of each error, the so-called least- squares solution.

By making m $(m > n)$ observations, we get m equations which we may write in matrix form as

$$
\begin{pmatrix} f_1 \\ f_2 \\ \vdots \\ f_m \end{pmatrix} =
\begin{pmatrix}
g_1^{(0)} & g_1^{(1)} & g_1^{(2)} & \cdots & g_1^{(n)} \\
g_2^{(0)} & g_2^{(1)} & g_2^{(2)} & \cdots & g_2^{(n)} \\
\vdots & \vdots & \vdots & & \vdots \\
g_m^{(0)} & g_m^{(1)} & g_m^{(2)} & \cdots & g_m^{(n)}
\end{pmatrix}
\begin{pmatrix} a_0 \\ a_1 \\ \vdots \\ a_n \end{pmatrix} +
\begin{pmatrix} e_1 \\ e_2 \\ \vdots \\ e_m \end{pmatrix} \tag{C.2}
$$

Where, since we not get an exact solution, we have included an error matrix e. Let us abbreviate this matrix equation to

$$f = Xa + e \tag{C.3}$$

To solve for a, we might think of multiplying Equation (C.3) by X^{-1}. Unfortunately, X is a non-square matrix and therefore cannot be inverted. We have

$$e = f - Xa \tag{C.4}$$

We form the sum of the square of each error by computing $e^T e$

$$e^T e = (f - Xa)^T (f - Xa) \tag{C.5}$$

$$= f^T f - f^T Xa - a^T X^T f + a^T X^T Xa$$

(For two matrices A and B we have $(AB)^T = B^T A^T$).

By taking the gradient of $(e^T e)$ with respect to a we find how the errors change as the coefficients change

$$\partial(e^T e)/\partial(a) = 0 - 2X^T f + 2X^T X a \tag{C.6}$$

where we have used the fact that

$$\frac{\partial}{\partial a}(f^T X a) = \frac{\partial}{\partial a}(a^T X^T f) = X^T f \tag{C.7a}$$

and

$$\frac{\partial}{\partial a}(a^T x^T X a) = 2X^T X a \tag{C.7b}$$

(Note that $(e^T e)$ is a scalar and that the transpose of a scalar is equal to itself)

The sum of the square of each error is minimized when setting the gradient equal to zero, thus

$$2X^T X a - 2X^T f = 0 \tag{C.8}$$

$$a = (X^T X)^{-1} X^T f$$

$(X^T X)^{-1} X^T$ is commonly referred to as the pseudo-inverse of X.

As an example, consider the case where a quadratic curve

$$f(x) = a_0 + a_1 x + a_2 x^2 \tag{C.9}$$

is fitted to N observation points (x_j, f_j) where $j = 1, 2, \ldots, N$. We then have

$$X = \begin{bmatrix} 1 & x_1 & x_1^2 \\ 1 & x_2 & x_2^2 \\ \vdots & \vdots & \vdots \\ 1 & x_N & x_N^2 \end{bmatrix} \quad X^T = \begin{bmatrix} 1 & 1 & \cdots & 1 \\ x_1 & x_2 & \cdots & x_N \\ x_1^2 & x_2^2 & \cdots & x_N^2 \end{bmatrix} \tag{C.10}$$

$$X^T X = \begin{bmatrix} N & s_1 & s_2 \\ s_1 & s_2 & s_3 \\ s_2 & s_3 & s_4 \end{bmatrix} \tag{C.11}$$

where

$$s_1 = \sum x_j \quad s_2 = \sum x_j^2 \quad s_3 = \sum x_j^3 \quad s_4 = \sum x_j^4 \tag{C.12}$$

$$(X^T X)^{-1} = \frac{1}{\det(X^T X)} \begin{bmatrix} (s_2 s_4 - s_3^2) & -(s_1 s_4 - s_3 s_2) & (s_1 s_3 - s_2^2) \\ -(s_1 s_4 - s_2 s_3) & (N s_4 - s_2^2) & -(N s_3 - s_1 s_2) \\ (s_1 s_3 - s_2^2) & -(N s_3 - s_2 s_1) & (N s_2 - s_1^2) \end{bmatrix} \tag{C.13}$$

$$X^T f = \begin{bmatrix} \sum f_j \\ \sum x_j f_j \\ \sum x_j^2 f_j \end{bmatrix} \tag{C.14}$$

This is a general result.

Now assume that $f(x)$ is the intensity distribution along a row of pixels and that we want to determine the position x_p of its maximum. By differentiating Equation (C.9) we find this position to be

$$x_p = -\frac{a_1}{2a_2} \tag{C.15}$$

For simplicity, let's assume that we fit the distribution to three points. (This is really not a good example since it gives an exactly determined system, but it is space-saving!) and that we have translated our coordinate system such that $x_1 = -1, x_2 = 0$ and $x_3 = 1$. Then we get $s_1 = 0, s_2 = 2, s_3 = 0$ and $s_4 = 2$, and

$$(X^T X)^{-1} = \frac{1}{\det(X^T X)} \begin{bmatrix} 4 & 0 & -4 \\ 0 & 2 & 0 \\ -4 & 0 & 6 \end{bmatrix} \tag{C.16}$$

$$X^T f = \begin{bmatrix} f_1 + f_2 + f_3 \\ -f_1 + f_3 \\ f_1 + f_3 \end{bmatrix} \tag{C.17}$$

This gives

$$\begin{bmatrix} a_0 \\ a_1 \\ a_2 \end{bmatrix} = \frac{1}{\det(X^T X)} \begin{bmatrix} 4f_2 \\ 2(-f_1 + f_3) \\ 2f_1 - 4f_2 + 2f_3 \end{bmatrix} \tag{C.18}$$

and

$$x_p = \frac{f_1 - f_3}{2f_1 - 4f_2 + 2f_3} \tag{C.19}$$

The procedure will then be to first determine the intensity maximum to the nearest pixel x_2 and find the intensities of the first pixels to the left and right (x_1 and x_3) and then find the position of the maximum with subpixel accuracy according to Equation (C.19).

Bibliography

1 BASICS

Further reading

Born, M., and Wolf, E. (1970) *Principles of Optics*, Pergamon Press, Oxford.
Hecht, E., and Zajac, A. (1974) *Optics*, Addison-Wesley, Reading, Massachusetts.
Jenkins, F. A., and White, H. E. (1957) *Fundamentals of Optics*, McGraw-Hill, New York.
Klein, M. V. (1970) *Optics*, Wiley, New York.
Longhurst, R. S. (1967) *Geometrical and Physical Optics*, Wiley, New York.
Meyer-Arendt, J. R. (1972) *Introduction to Classical and Modern Optics*, Prentice-Hall, Englewood Cliffs, New Jersey.
Stone, I. M. (1963) *Radiation and Optics*, McGraw-Hill, New York.
Young, H. D. (1968) *Fundamentals of Optics and Modern Physics*, McGraw-Hill, New York.

2 GAUSSIAN OPTICS

Further reading

Hecht, E., and Zajac, A. (1974) *Optics*, Addison-Wesley, Reading, Massachusetts.
Klein, M. V. (1970) *Optics*, Wiley, New York.

3 INTERFERENCE

References and further reading

Baldwin, R. G., Gordon, G. B. and Rudè A. F. (1971) *Hewlett-Packard Journal*, December 14.
Cook, A. H. (1971) *Interference of Electromagnetic Waves*, Clarendon Press, Oxford.
Dyson, J. (1970) *Interferometry as a Measuring Tool*, The Machinery Publishing Co., Brighton.
Francon, M., and Mallick, S. (1971) *Polarization Interferometers*, Wiley-Interscience, London.
Gaal, P., Jani, P., and Czitrovszky, A. (1993). Opt. Engineering, **32**, 2574–77.
Hecht, E., and Zajac, A. (1974), *Optics*, Addison-Wesley, Reading, Massachusetts.
Klein, M. V. (1970) *Optics*, Wiley, New York.
Steel, W. H. (1984) Interferometry, 2nd edition, University Press, Cambridge.

4 DIFFRACTION

Further reading

Goodman, J. W. (1968) *Introduction to Fourier Optics*, McGraw-Hill, New York.
Hecht, E., and Zajac, A. (1974) *Optics*, Addison-Wesley, Reading, Massachusetts.
Klein, M. V. (1970) *Optics*, Wiley, New York.

5   LIGHT SOURCES AND DETECTORS

References

Barbe, D. F., and Campana, S. B. (1977) Imaging array using the charge-coupled concept, in Kazan, B. (Ed.) *Advances in Image Pickup and Display*, Academic Press, New York.
Barbe, D. F. (1975) Proc. IEEE, **63**, 38–67.
Barbe, D. F., and Campana, S. B. (1976) Aliasing and MTF effects in photosensor arrays, in Jespers, P. G., van de Wiele, F., and White, M. H. (eds.) *Solid State Imaging*, Noordhoff, Leyden.
Biberman, L. M., and Nudelman, S. (1971) (eds.) Photoelectric imaging devices, Vols. I and II, Plenum Press, New York.
Boyle, W. S., and Smith, G. E. (1970) *Bell System Tech. J.*, **49**, 587–93.
Goodman, J. W. (1968) *Introduction to Fourier Optics*, McGraw-Hill, New York.
Grob, B. (1984) *Basic Television and Video Systems*, 5th Edition, McGraw-Hill, New York.
Hecht, J. (1992) *The Laser Guidebook*, 2nd Edition, McGraw-Hill, New York.
Javan, A., Bennett Jr., W. R., and Herriot, D. R. (1961) Phys. Rev. Lett., **6**, 106–10.
Klein, M. V. (1970) *Optics*, Wiley, New York.
Longhurst, R. S. (1967) *Geometrical and Physical Optics*, Wiley, New York.
Maiman, H. (1960) Nature, **187**, 493–4.
Oriel Corporation (1994) *Light Sources. Monochromators. Detection Systems*, Stratford, Connecticut.
Slater, P. N. (1980) *Remote Sensing. Optics and Optical Systems*, Addison-Wesley, Reading, Massachusetts.
Urbach, J. C., and Meier, R. W. (1966) *Applied Optics*, **5**, 666–67.
Van Wezel, R. in King, G. J. (Ed.) (1981) *Video Handbook*, Newnes Technical Books, London.
Vest, C. M. (1979) *Holographic Interferometry*, Wiley, New York.

6   HOLOGRAPHY

References

Abramson, N. (1970) *Applied Optics*, **9**, 97–101.
Abramson, N. (1972) *Applied Optics*, **11**, 1143–47.
Abramson, N. (1974) *Applied Optics*, **13**, 2019–25.
Abramson, N. (1981) *The Making and Evaluation of Holograms*, Academic Press, London.
Aleksoff, C. C. (1971) *Applied Optics*, **10**, 1329–41.
Aleksoff, C. C. (1974) in Erf.
Boone, P. M. (1974) *Digest of the 1974 IEEE International Optical Computing Conference*, Zürich, April 9–11.
Boone, P. M. (1975) *Optica Acta*, **22**, 579–89.
Burch, J. M. (1965) *Production Engineering*, **44**, 431–42.
Butters, J. N., and Leendertz, J. A. (1977) *Measurement and Control*, **4**, 349–54.
Caulfield, H. J. (1979) *Handbook of Optical Holography*, Academic Press, New York.
Collier, R. J., Burckhardt, C. B., and Lin, L. H. (1971) *Optical Holography*, Academic Press, New York.
Dandliker, R. and Thalmann, (1985) *Optical Engineering*, **24**, 824–31.
Erf, R. (Ed.) (1974) *Holographic Nondestructive Testing*, Academic Press, New York.
Gabor, D. (1949) *Proc. Roy. Soc.* **A197**, 454–87.
Haines, K. A. and Hildebrand, B. P. (1965) *Physics Letters*, **19**, 10–1.
Hariharan, P. (1984) *Optical Holography*, Cambridge University Press, Cambridge.
Hariharan, P., Ramanathan, C. S., and Kaushik, G. S. (1971) *Opt. Commun.*, **3**, 246–47.

Høgmoen, K. and Løkberg, O. J. (1976) *Applied Optics*, **16**, 1869–75.
Horman, M. H. (1965) *Applied Optics*, **4**, 333–6.
Jones, R., and Wykes, C. (1989) *Holographic and Speckle Interferometry*, Cambridge University Press, Cambridge.
Leith, E. N., and Upatnieks, J. (1962) *JOSA*, **52**, 1123–30.
Løkberg, O. J. (1979) *Applied Optics*, **18**, 2377–84.
Løkberg, O. J. (1980) *Phys. Technol.*, **11**, 16–22.
Løkberg, O. J., Høgmoen, K., and Holje, O. M. (1979) *Applied Optics.*, **18**, 763–5.
Lokberg, O. J., and Slettemoen, G. (1987) in Wyant, J. C. and Shannon, R. (Eds.), *Applied Optics and Optical Engineering*, Vol. X, Academic Press, New York.
Macovski, A., Ramsey, S. D., and Schaefer, L. F. (1971) *Applied Optics*, **10**, 2722–7.
Malmo, J. T., and Vikhagen, E. (1988) *Experimental Technology*, **12**, 28–30.
Metherell, A. F., Spinak, S., and Pisa, E. J. (1969) JOSA, **59**, 1534.
Molin, N.-E., and Stetson, K. A. (1971) *Optik*, **33**, 399–422.
Powell, R. L., and Stetson, K. A. (1965) JOSA, **55**, 1593–8.
Schwomma, O. (1972) Austrian patent No. 298830.
Smith, H. M., *Principles of Holography*, Wiley-Interscience, New York.
Vest, C. M. (1979) *Holographic Interferometry*, Wiley, New York.
Walles, S. (1969) *Ark. Fys.*, **40**, 299–403.
Wilson, A. D., Lee, C. H., Lominac, H. R., and Strope, D. H. (1971) *Exp. Mech.*, **11**, 1–6.

7 MOIRÉ TECHNIQUE

References

Asundi, A. K., Chan, C. S., and Sajan, M. R. (1994) Optical Engineering, **33**, 2760–69.
Brooks, R. E., and Heflinger, L. O. (1969) *Applied Optics*, **8**, 935–39.
Cardenas- Garcia, J. F., Zheng, S., and Shen, F. Z. (1994) Optics and Lasers in Engineering, **21**, 77–98.
Cline, H. E., Lorensen, W. E., and Holik, A. S. (1984) *Applied Optics*, **23**, 1454–59.
Dessus, B., and Leblanc, M. (1973) *Opto-Electronics*, **5**, 369–91.
Doty, J. L. (1983) *JOSA*, **73**, 366–72.
Gåsvik, K. J. (1983) *Applied Optics*, **22**, 3543–8.
Gåsvik, K. J., and Fourney, M. E. (1986) *J. Appl. Mech.*, **53**, 652–6.
Halioua, M., Krishnamurthy, R. S., Liu, H., and Chiang, F. P. (1983) *Applied Optics*, **22**, 850–55.
Harding, K. G., and Harris, J. S. (1983) *Applied Optics*, **22**, 856–61.
Hazell, C. R., and Niven, R. D. (1968) *Exp. Mech.*, **8**, 225–31.
Idesawa, M., Yatagai, T., and Soma, T. (1977) *Applied Optics*, **16**, 2152–62.
Lichtenberg, F. K. (1955) *Proc. SESA*, **12**, 83–98.
Liasi, E., and North, W. P. T. (1994) Optical Engineering, **33**, 1200–05.
Pirodda, L. (1982) *Opt. Engineering*, **21**, 640–9.
Post, D. (1971) *Applied Optics*, **10**, 901–07.
Post, D. (1982) *Opt. Engineering*, **21**, 458–67.
Reid, G. T., Rixon, R. C., and Messer, H. I. (1984) *Opt. and Laser Technology*, **16**, 315–19.
Sciammarella, C. A. (1972) *J. Phys. E: Scientific Instruments*, **5**, 833–45.
Sciammarella, C. A. (1982) *Exp. Mech.*, **22**, 418–33.
Takasaki, H. (1973) *Applied Optics*, **12**, 845–50.
Takasaki, H. (1982) *Opt. and Lasers in Engineering*, **3**, 3–14.
Takeda, M. (1982) *Opt. and Lasers in Engineering*, **3**, 45–52.
Varman, P. O. (1984) *Opt. and Lasers in Engineering*, **5**, 41–58.
Vest, C. M., and Sweeney, D. W. (1972) *Applied Optics*, **11**, 449–54.
Yatagai, T., and Idesawa, M. (1982) *Opt. and Lasers in Engineering*, **3**, 73–83.

8 SPECKLE METHODS

References

Archbold, E., and Ennos, A. E. (1972) *Optica Acta*, **19**, 253–71.
Archbold, E., Burch, J. M., and Ennos, A. E. (1970) *Optica Acta*, **17**, 883–98.
Burch, J. M., and Forno, C. (1975) *Optical Engineering*, **14**, 178–85.
Burch, J. M., and Tokarski, J. M. (1968) *J. Optica Acta*, **15**, 101–11.
Butters, J. N., and Leendertz, J. A. (1971) *J. Phys. E: Scientific Instruments*, **4**, 277–79.
Cloud, C. (1975) *Applied Optics*, **14**, 878–84.
Dainty, J. C. (Ed.) (1975) *Topics in Applied Physics 9: Laser Speckle and Related Phenomena*, Springer Verlag, Berlin.
Duffy, D. E. (1972) *Applied Optics*, **11**, 1778–81.
Duffy, D. E. (1974) *Exp. Mech.*, **14**, 378–84.
Erf, R. K. (Ed.) (1978) *Speckle Metrology*, Academic Press, New York.
Forno, C. (1975) *Optics and Laser Technology*, **16**, 217–21.
Fourney, M. E. (1978) in Erf (1975).
Goodman, J. W. (1975) *Statistical Properties of Laser Speckle Patterns*, in Dainty (1975).
Groh, G. (1968) in Robertson, E. R., and Harvey, (Eds), *The Engineering Uses of Holography*, Cambridge University Press, Cambridge.
Halliwell, N. A., and Pickering, C. J. (1993) Chapter 7 in Robinson, D. W., and Reid, G. T. (Eds), *Interferogram Analysis. Digital Fringe Pattern Measurement Techniques*, Institute of Physics Publishing, Bristol.
Hung, Y. Y. (1974) *Opt. Commun.*, **11**, 732–35.
Hung, Y. Y. (1978) in Erf (1978).
Hung, Y. Y. (1982) *Optical Engineering*, **21**, 391–5.
Hung, Y. Y., and Hovanesian, J. D. (1972) *Exp. Mech.*, **12**, 454–60.
Hung, Y. Y., and Taylor, C. E. (1973) *Soc. Photo-Opt. Instrum. Eng.*, **14**, 169–75.
Hung, Y. Y., Daniel, I. M., and Rowlands, R. E. (1975) *Applied Optics*, **14**, 618–60.
Jones, R., and Wykes, C. (1989) *Holographic and Speckle Interferometry*, Cambridge University Press, Cambridge.
Leendertz, J. A. (1979) *J. Phys. E: Scientific Instruments*, **3**, 214–18.
Leendertz, J. A., and Butters, J. N. (1973) *J. Phys. E: Scientific Instruments*, **6**, 1107–10.
Mohan, N. K., Saldner, H. O., and Molin, N.-E. (1993) Optics Letters, **18**, 1861–3.
Saldner, H. O. (1994) *Electronic holography and shearography in experimental mechanics*. Licentiate thesis, Luleå University of Technology.
Tiziani, H. J. (1972) *Opt. Commun.*, **5**, 271–6.

9 PHOTOELASTICITY AND POLARIZED LIGHT

Further reading

Clarke, D., and Graininger, J. F. (1971) *Polarized Light and Optical Measurement*, Pergamon Press, Oxford.
Fourney, M. E. (1968) *Exp. Mech.*, **8**, 33–38.
Frocht, M. M. (1961) *Photoelasticity*, Vols. I and II, Wiley, New York.
Gåsvik, K. J. (1976) *Exp. Mech.*, **16**, 146–50.
Hecht, E., and Zajac, A. (1974) *Optics*, Addison-Wesley, Reading, Massachusetts.
Klein, M. V. (1970) *Optics*, Wiley New York.
Kuske, A., and Robertson, G. (1974) *Photoelastic Stress Analysis*, Wiley, London.
Simmons, J. W., and Guttmann, M. J. (1970) *States Waves and Photons: A Modern Introduction to Light*, Addison-Wesley, Reading, Massachusetts.

Theocaris, P. S., and Gdoutos, E. E. (1979) *Matrix Theory of Photoelasticity*, Springer-Verlag, Berlin.

10 DIGITAL IMAGE PROCESSING

References

Baxes, G. A. (1984) *Digital Image Processing. A practical Primer*, Prentice Hall, New York.

Canny, J. (1986) *IEEE Transactions on Pattern Analysis and Machine Intelligence*, **PAMI-8**, 679–98.

Faig, W. (1975) *Photogram. Eng. Remote Sensing*, **41**, 1479–86.

Fram, J. R., and Deutsch, E. S. (1975) *IEEE Trans. Computers*, **C-24** (6), 616–28.

Gonzales, R. C., and Wintz, P. (1987) *Digital Image Processing*, 2nd Edition, Addison Wesley, Reading, Massachusetts.

Lenz, R. K., and Tsai, R. Y. (1988) *IEEE Trans. Pattern Anal. Mach. Intell.*, **10**, 713–20.

Marr, D., and Hildreth, E. *Proc. Roy. Soc. of London*, **B207**, 187–217.

Niblack, W. (1988) *An Introduction to Digital Image Processing*, Prentice Hall, New York.

Pratt, W. K. (1991) *Digital Image Processing*, 2nd Edition, Wiley New York.

Prewitt, J. M. S. (1970) in Lipkin, B. S. and Rosenfeld, A. (Eds.). *Picture Processing and Psycho-pictorics*, Academic Press, New York.

Roberts, L. G. (1965) in Tippett, J. T. *et al.* (Eds.), *Optical and Electro-optical Information Processing*, MIT Press, Cambridge, Massachusetts.

Rosenfeld, A., and Kak, A. C. (1982) *Digital Picture Processing*, 2nd Edition, Academic Press, New York.

Shih, S.-W., S., Hung, Y.-P., and Lin, W.-S. (1993) *Opt. Engineering*, **32**, 138–49.

Tsai, R. Y. (1987) *IEEE J Robot. Automat.*, **RA-3**, 323–44.

Vernon, D. (1991) *Machine Vision*, Prentice Hall, New York.

11 FRINGE ANALYSIS

References

Bareket, N. (1985) *Proc. SPIE*, **551**, 12–7.

Bone, D. J., Bachor, H. A., and Sandeman, R. J. (1986) *Applied Optics*, **25**, 1653–60.

Button, B. L., Cutts, J., Dobbins, B. N., Moxon, C. J., and Wykes, C. (1985) *Optics and Laser Technology*, **17**, 189–92.

Carré, P. (1966) *Metrologia*, **2**, 13–23.

Chang, M., Lin, P. P., and Tai, W. C. (1994) Optics and Lasers in Engineering, **20**, 163–76.

Choudry, A. (1981) *Applied Optics*, **20**, 1240–4.

Creath, K. (1986) *Proc. SPIE*, **680**, 19–28.

Creath, K. (1993) Chapter 4 in Robinson and Reid

Frankowski, G., Stobbe, I., Tischer, W., and Schillke, F. (1989) *Proc. SPIE*, **1121**, 89–100.

Gåsvik, K. J., and Robbersmyr, K. G. (1994) *Opt. Engineering*, **33**, 246–50.

Gåsvik, K. J., Robbersmyr, K. G., and Vadseth, T. (1989) *Proc. SPIE*, **1163**, 64–70.

Ghiglia, D. C., Mastin, G. A., and Romero, L. A. (1987) *JOSA* (A), **4**, 267–80.

Ghiglia, D. C., and Romero, L. A. (1994) JOSA (A), **11**, 107–17.

Gierloff, J. J. (1987) *Proc. SPIE*, 818.

Greivenkamp, J. E. (1984) *Optical Engineering*, **23**, 350–52.

Halliwell, N. A., and Pickering, C. J. (1993) Chapter 7 in Robinson and Reid.

Harding, K. G., Coletta, M. P., Van Dommelen, (1988) *Proc. SPIE*, **1005**, 169–78.

Hariharan, P., Oreb, B. F., and Eiju, T. (1987) *Applied Optics*, **26**, 2504–05.

Hovanesian, J. Der, and Hung, Y. Y. (1982) *Proc. SPIE*, **360**, 88–9.

Huntley, J. M. (1989) *Applied Optics Letters*, **28**, 3268–70.

Ichioka, Y., and Inuiya, M. (1972) *Applied Optics*, **11**, 1507–14

Jin, G., Bao, N., and Chung, P. S. (1994) Optical Engineering, **33**, 2733–37.

Joenathan, C., (1994) Applied Optics, **33**, 4147–55.

Judge, T. R., and Bryanston-Cross, P. J. (1993) Optics and Lasers in Engineering, **21**, 199–239.

Kujawinska, M. (1993) Chapter 5 in Robinson and Reid.

Kujawinska, M., and Robinson, D. W. (1988) *Applied Optics*, **27**, 312–20.

Kujawinska, M., Spik, A., and Robinson, D. W. (1989) *Proc. FASIG 'Fringe Analysis' 89'*, Loughborough, UK 4–5 April.

Kujawinska, M. (1993b) Optics and Lasers in Engineering, **19**, 261–68.

Kwon, O. Y., and Shough, D. M. (1985) *Proc. SPIE*, **599**, 273–9.

Kwon, O. Y., Shough, D. M., and Williams, R. A. ((1987) *Optics Letters*, **12**, 855–7.

Larkin, K. G., and Oreb, B. F. (1992) *JOSA* (A) **9**, 1740–8

Lassahn, G. D., Lassahn, J. K., Taylor, P. L., and Deason, V. A. (1994) Optical Engineering, **33**, 2039–44.

Malcolm, A. A., Burton, D. R., and Lalor, M. J. (1989) *Proc. FASIG 'Fringe Analysis' 89'*, Loughborough, UK, April.

Magee, E. P., and Welsh, B. M. (1994) Optical Engineering, **33**, 3810–17.

Mertz, L. (1983) *Applied Optics*, **22**, 1535–9.

Morgan, C. J. (1982) *Optics Letters*, **7**, 368–70

Morimoto, Y., and Fujisawa, M. (1994) Optical Engineering, **33**, 3709–14.

Morimoto, Y., Gascoigne, H. E., and Post, D. (1994) Optical Engineering, **33**, 2646–53.

Nugent, K. A. (1985) *Applied Optics*, **24**, 3101–5.

Perry Jr., K. E., and McKelvie, J. (1993) Optics and Lasers in Engineering, **19**, 269–84.

Quan, C., Bryanston-Cross, P. J., and Judge, T. R. (1993) Optics and Lasers in Engineering, **18**, 79–108.

Robinson, D. W. (1983) *Applied Optics*, **22**, 2169–76.

Robinson, D. W. (1993) Chap 6 in Robinson and Reid.

Robinson, D. W., and Reid, G. T. (Eds.), (1993) *Interferogram Analysis. Digital Fringe Pattern Measurement Techniques*, Institute of Physics Publishing, Bristol.

Robinson, D. W., and Williams, D. C. (1986) *Optics Communications*, **57**, 26–30.

Rosvold, G. O. (1990) *Applied Optics*, **29**, 237–41.

Schorner, J, *et al.* (1991) *Optics and Lasers in Engineering*, **14**, 283–92.

Schwider, J., Burow, R., Elssner, K.–E., Grzanna, J., Spolaczyk, R., and Merkel, K. (1983) *Applied Optics*, **22**, 3421–32.

Schwider, J., Falkenstørfer, O., Schreiber, H., Zøller, A., and Streibl, N. (1993) Optical Engineering, **32**, 1883–85.

Simova, E. S., and Stoev, K. N. (1993) Optical Engineering, **32**, 2286–94.

Singh, H., and Sirkis, J. S. (1994) Applied Optics, **33**, 5016–20.

Smythe, R., Moore, R. (1984) *Optical Engineering*, **23**, 361–4.

Stephenson, P., Burton, D. R., and Lalor, M. J. (1994) Optical Engineering, **33**, 3703–08.

Surrel, Y. (1993) Applied Optics, **33**, 3598–600.

Takeda, M., and Mutoh, K. (1983) *Applied Optics*, **22**, 3977–82.

Takeda, M., Ina, H., and Kobayashi, S. (1982) *JOSA*, **72**, 156–60.

Towers, D. P., Judge, T. R., and Bryanston-Cross, P. J. (1991) *Optics and Lasers in Engineering*, **14**, 239–82.

Vikhagen, E. (1990) *Applied Optics*, **29**, 137–44.

Vrooman, H. A., and Maas, A. A. (1989) *Proc. FASIG 'Fringe Analysis'89'*, Loughborough UK, 4–5 April.

Vikram, C. S., Witherow, W. K., and Trolinger, J. D. (1993) Applied Optics, **32**, 6250–52.

Womack, K. H. (1984) *Optical Engineering*, **23**, 391–5.

Wyant, C. J. (1982) Laser Focus *May* 65–71.

Wyant, J. C., Koliopoulos, C. L., Bhushan, B., and George, O. E. (1984) *ASLE Trans*, **27**, 101–13.

Yatagai, T., Idesawa, M., Yamaashi, Y., and Suzuki, M. (1982) *Optical Engineering*, **21**, 901–6.

Yatagai, T. (1993) Chapter 3 in Robinson and Reid.

Yoshizawa, T., and Tomisawa, T. (1993) Optical Engineering, **32**, 1668–74.

Yu, Q., Andresen, K., Osten, W., and Jüptner, W. P. O. (1994) Optical Engineering, **33**, 1630–37.

12   SOME OTHER OPTICAL METROLOGY TECHNIQUES

References

Beckmann, P., and Spizzichino, A. (1963) *The Scattering of Electromagnetic Waves from Rough Surfaces*, Pergamon Press, Oxford.

Berthold, III, J. W. (1993) Status and review of fiber optic sensors in industry, p. 271 in Udd. (1993)

Cook, R. O., and Hamm, C. W. (1979) *Applied Optics*, **18**, 3230–82.

Culshaw, B. (1956) *J. Opt. Sensors*, **1**, 327–000.

Durst, F., Melling, A., and Whitelaw, J. H. (1981) *Principles and Practice of Laser-Doppler Anemometry*, Academic Press, London.

Gåsvik, K. J. (1980) *Optica Acta*, **27**, 965–80.

Gåsvik, K. J. (1981) *Optica Acta*, **28**, 131–38.

Keiser, G. (1991) *Optical Fiber Communications*, 2nd Edition, McGraw-Hill, New York.

Kyuma, K., Tai, S., and Nunoshita, M. (1982) *Opt. and Lasers in Engineering*, **3**, 155–82.

Leonhardt, K., Rippert, K. H., and Tiziani, H. J. (1988) SPIE vol. 1009, 22–29.

Neal, W. E. J., and Fane, R. W. (1973) *J Phys. E.*, **6**, 409–16.

Palais, J. C. (1992) *Fiber Optic Communications*, Prentice- Hall, Englewood Cliffs, New Jersey.

Passaglia, E. (Ed.) (1964) *Ellipsometry in the Measurement of Surfaces and Their Films*, National Bureau of Standards, Washington.

Pedersen, H. M. (1976) *JOSA*, **66**, 1204–10.

Persson, U. (1992) *Opt. and Lasers in Engineering*, **17**, 61–7.

Philips, G. J. (1980) *Mech Eng. July*, 28–00.

Schiffer, R. (1987) Journal of Modern Optics, **34**, 1145–68.

Senior, J. M. (1985) *Optical Fiber Communications. Principles and Practice*, Prentice Hall, Englewood Cliffs, New Jersey.

Udd, E. (Ed.) (1991) *Fiber Optic Sensors: An Introduction for Engineers and Scientists*, Wiley, New York.

Udd, E. (1993) *Fiber Optic Sensors*, SPIE Critical Reviews CR44.

Yu, F. T. S., and Khoo, I. C. (1990) *Principles of Optical Engineering*, Chapter 10, Wiley, New York.

Index